# ROLES OF TECHNOLOGY DURING COVID-19

# ROLES OF TECHNOLOGY DURING COVID-19

**Authors**

Kanishtrayen Baskaran
Swati Anant
Nicholas Hamzea
Annie Li

**Editors**

Austin Mardon
Catherine Mardon

**Graphic Designer**

Amy Zhao

Golden Meteorite Press

Published 2020

First Printing: 2020

ISBN: 978-1-77369-177-0

Golden Meteorite Press
103 11919 82 St NW
Edmonton, AB T5B 2W3
www.goldenmeteoritepress.com


We acknowledge the support of Canada Service Corps, TakingITGlobal and the Government of Canada in promotional materials associated with the Project. Thank you for your support.

# Table of Contents

# Acknowledgements

We acknowledge the support of Canada Service Corps, TakingITGlobal and the Government of Canada in promotional materials associated with the Project.

Thank you for your support.

# Chapter 1: Introduction

The COVID-19 pandemic has had devastating impacts on every functional aspect of society, with hundreds of thousands dead and millions more infected in just a few short months. Although many countries and regions, such as Canada and most of Europe, have shown significant decreases in the number of active cases, the pandemic is far from over and, on a global level, the peak has yet to even be reached as of July 2020 (Neustaeter, 2020). What started as a small outbreak in Wuhan, China back in late 2019 quickly snowballed into a "public health emergency of international concern" by the end of January before being declared a global pandemic on March 11, 2020 by the World Health Organization (WHO, 2020a). Within an instant, countries were scrambling to close borders, shut down public spaces and workplaces, and draft new regulations to reduce the spread of COVID-19; the world was in a state of limbo and it seemed as if nobody knew what to do next. For many, this was their first time experiencing a global crisis to this degree and concerns from citizens and governments alike rapidly formed. What will happen to the economy? How will children receive an education? When can families reunite once again? These are just some of the questions that dominated citizens as this "new normal" began to set in. Despite taking some time to adjust, however, solutions to almost all of these concerns began to be put in place largely because of advancements in modern technology.

Modern technology has arguably been one of the greatest strengths and weaknesses in the fight against COVID-19. It is also one of the key distinguishing factors between COVID-19 and pandemics of the past, as the advancement of technology over the last several decades has led to unforeseen benefits and consequences. It has impacted everything ranging from how COVID-19 is detected, diagnosed, and treated, to socio-political and economic changes that have occurred as a result of this pandemic.

Perhaps the greatest asset in the fight against this pandemic has been the ability to use existing technologies to detect, diagnose, and treat COVID-19 while also directly protecting individuals from the virus. The presence of infectious diseases in the healthcare system is by no means a new concept, and the need for sterility and protection in such a setting has led to masks that can block the transmission of pathogens, specifically surgical and N95 masks that had been developed decades prior. In fact, the N95 masks that protect healthcare workers today were invented back in 1995 (Page, 2020). With the COVID-19 pandemic being as widespread as it is, the technology used in not only developing the masks but also increasing the rate of their production has been invaluable, as they, along with other forms of personal protective equipment (PPE), have proven to be some of the most simple yet effective tools in reducing the transmission of airborne pathogens. Furthermore, diagnostic technologies such as antibody tests and immunohistochemistry have changed the efficiency of testing. The progress of immunological research and modern technologies have allowed the genome of the novel coronavirus to be determined (Science Daily, 2020), and additional treatment strategies to be tested at a higher rate than in the past. Additionally, advances in vaccine technology as a result of genetic engineering have allowed us to develop more efficacious vaccines.

In addition to the application of pre-existing technologies, this pandemic has provided an opportunity for human ingenuity and innovation to shine, with many new technologies being used and developed to curb the spread of COVID-19. This includes the use of artificial intelligence, 3D printing to produce PPE and other Equipment, and even apps that the general public can download. One example of this is the implementation of the "COVID Alert" app by the government of Canada which uses bluetooth to detect the phones of individuals that a person has come into close contact with and will then send them an alert if any of their contacts has tested positive. This technology does not collect the location, name, or any other personal information, and has the potential to significantly boost the efficiency of contact tracing on a large scale. New developments such as these will continue to help the fight against COVID-19 and may even prove to be a valuable tool for future emergencies (Aiello, 2020).

Technology has also had a substantial impact on our health from an individual and public health perspective, thus having profound societal implications. The development of social media and online meeting platforms have allowed people to connect with one another like never before. In a time such as this, where social distancing measures are strict and widespread, feelings of isolation would normally be incredibly common which can have numerous negative mental and physical health effects. This is especially true for the senior population, who due to their increased risk of developing complications as a result of COVID-19, must remain especially physically isolated from the general population. However, many platforms such as Zoom, FaceTime, and Google Meets are being used to ensure that people are able to connect and speak with one another, thus reducing the potential mental health consequences of this pandemic. At the same time, it is possible that reliance on technology has dire consequences with respect to mental health. With regard to physical health, many gyms and training facilities are now transitioning towards the use of technology to deliver their services virtually. This may be through uploading pre-recorded workouts and workout schedules, hosting live fitness classes, or allowing people to book appointments for in-person sessions to ensure that social distancing measures within facilities may be met.

Despite its benefit in the fight against COVID-19, however, it has become one of society's greatest vulnerabilities. This is because the demand for and dependence on technology by society as a whole leads to significant consequences when those needs are unable to be met. One of the most critical examples of this is the increased risk faced by healthcare workers due to an inadequate PPE supply chain which has led to shortages. In fact, many workers, such as those working at two

Toronto hospitals, are being told that they will have to reuse decontaminated N95 masks, which are meant to be single use only, as a result of the shortage (Warren, 2020). In addition to this, shortages of ventilators and other critical tools in healthcare settings have unfortunately led to the death of many people, as was the case in Italy in April when doctors were put into a position of deciding who lives and who dies due to the shortage in Intensive Care Unit (ICU) beds and equipment (Beall, 2020). Needless to say, the significantly increased demand for technology has been one of the most significant bottlenecks in the treatment of COVID-19 patients. Although the situation has improved substantially, many workers are still required to reuse PPE for prolonged periods of time which poses a potential danger to both the healthcare provider and the patient.

Despite the scientific nature of the advancements that have been made with regards to technology, there have been profound related impacts in all social aspects of society, including culture, politics, and economics. With the pandemic confining everyone to their own home and preventing people from seeing one another face-to-face, the incorporation of new technology such as video calling platforms to allow some level of connection was seen as a necessity by many. This necessity also extends to the workplace, with many businesses switching to teleconferencing softwares and a "work from home" model in order to adapt to the changes in lifestyle brought about by the pandemic. However, while some businesses have been able to adapt and thus resume operation, many businesses have still seen dangerous cuts to revenue or have gone bankrupt. The result of this has been one of the worst economic downturns in human history that will have a plethora of long-lasting and significant consequences. With regards to politics, many new policies have been implemented to curb the spread of COVID-19, such as the closing of businesses and public settings, which as mentioned have their own consequences, the mandating of masks, and the closure of borders. Many of the decisions pertaining to each of these social aspects were largely impacted by the presence of technology, clearly demonstrating that its reach is far greater than one might think.

One of the key downfalls of technology throughout the COVID-19 pandemic is the role that the internet and media have played in the perpetuation of misinformation, thus significantly harming the fight against COVID-19. In a world of close to 8 billion people, approximately 3.5 billion people are online in some way, with the number of facebook users alone growing from 518 million in 2010, to a staggering 2.26 billion in 2018 (Ortiz-Ospina, 2019), and the number of new users joining social media platforms every year is by no means slowing down. This acts as a distinguishing factor between COVID-19 and previous pandemics, as the level at which information, and subsequently misinformation, can be created and spread is far greater than ever witnessed before. Unfortunately, misinformation is not only circulating throughout the internet, but is in fact being interacted with far more than credible information. The result of this is an increase in the spread of COVID-19, and people opposing vaccines and refusing to listen to public health advice. Although social media may be one of the greatest tools of the twenty-first century to keep individuals connected, it has arguably been one of society's greatest weaknesses in the fight against this virus.

A global event of this size and severity will certainly have numerous long-lasting impacts even after the virus itself has subsided. With billions of lives being directly affected and the world facing one of the worst economic downturns in history, it will take years to recuperate to a level that is close to the normal everyone once knew. Many of the implications and changes, however, will not necessarily directly impact citizens on an individual level. These impacts include changes to the economy and efficiency focused global supply chains, new pandemic and emergency preparedness strategies and policies, and even the hindrance of a transition towards a more sustainable future. On an individual level, people may begin to see changes to the workplace, travel, culture, and how society views the ill. Despite society eventually returning to a state that will be rather similar to what it was before, this pandemic will ultimately leave its mark in history and embed itself in daily life for the years to come.

With the advancement of technology being one of the key distinguishing factors between COVID-19 and past pandemics, it is crucial to understand the diverse and significant impacts that it has had on the fight against the virus. This book will take a multifaceted approach to understanding the implications of technological advancements in recent times, and will cover topics ranging from the use of technology to prevent, diagnose, and treat COVID-19, to its socioeconomic and political consequences.

The hope of this book is to provide an accurate portrayal of the role of technology during the COVID-19 pandemic. It will assess the impact of technology on various stakeholders such as individuals, society, businesses and the government. Furthemore, it will analyze the advancement of technology throughout COVID-19 and finally how technology has influenced change in everyday life to different degrees.

Research conducted on this topic will outline how technology has impacted various parties and further evaluate the benefits and detriments of its use. These conclusions will include findings pertaining to technology and its influence on the well-being of individuals and communities, as well as its impacts on a larger scale with regards to different levels of governments and businesses. Given the unique circumstances of the pandemic, we hope to inform readers and provide a changed perspective surrounding the role of technology. This information may potentially be utilized to make future educated decisions regarding policies in governments, businesses, or organizations as well as personal choices. The hope is that the aggregate of this research will aid in preventing the spread of misinformation surrounding the current pandemic and help lead to a more informed society.

# Chapter 2: Diagnostic Technologies

The COVID-19 pandemic is one of the many events in history which has instilled a great deal of fear and confusion among the general population. This nation have gone from seeing a single cough being as insignificant as a mosquito bite, to fearing them as if they were life-threatening murder hornets. One of the major points of speculation during the pandemic across the globe is surrounding testing. When should a person get tested? How accurate are current diagnostic technologies? As of July 2020, the general outlook is that anyone who has come in contact with a COVID-19 patient or is experiencing any mild flu-like symptoms should get tested. However, the actual mechanisms of testing have raised a great deal of uncertainty. According to an article on Global News, the main COVID-19 testing kit being utilized generates false negatives at least 20% of the time. Sources like this have generated fear and anxiety among people who have displayed symptoms and have not been able to socially distance, thus perpetrating the spread of this virus. So, what's the real story with regards to these tests? What do they even do, and how have the diagnostic technologies evolved over time?

## Determining the SARS-CoV-2 Genome

Before looking into the current testing methods for COVID-19, it is necessary to go back in time to the research event that allowed current testing strategies to be developed. This event was the detection and sequencing of the novel coronavirus genome. This study was published in the Chinese Medical Journal, and the rapidity of this discovery signals how recent technological advances may be lifesaving. In this study, bronchoalveolar lavage fluid samples were collected from the five initial patients that were hospitalized due to this virus in 2019 (Yasinski, 2020). Nucleic acids were extracted from these samples and a sequencing library was constructed by a transposase-based methodology using an Illumina sequence platform. The sequences were confirmed using Sanger sequencing. Illumina dye sequencing works by amplifying, sequencing, and analyzing the DNA, and a similar process is followed with Sanger sequencing. Both of these technologies, and many more, allowed for the identification of the novel coronavirus genome within about 3 days (Yasinski, 2020). This presents a stark contrast to the discovery of the SARS and MERS genomes, which took several months to identify. The identification of the genome has allowed for testing methods to be rapidly developed (Aguiar. et al,2020).

**Types of Diagnostic Technologies**

This examination of the diagnostic technologies present before and during COVID-19 will begin with the basics. There are two main commercial types of testing methods for the novel coronavirus – nucleic acid-based testing (NAT) and serology-based testing. NAT works through detection of RNA strands specific to the SARS-CoV-2 virus. It detects if an individual is currently infected with COVID-19. In NAT, nasopharyngeal swabs are used to sample respiratory secretions. The test extracts RNA from the sample, converts it to DNA and amplifies it using PCR with SARS-CoV-2 specific primers. An example of a NAT is the COVID-19 RT-PCR test, developed by LabCorb in 2020. RT-PCR is a real-time reverse transcription polymerase chain reaction test to detect nucleic acids from SARS-CoV-2 (Wu et al, 2020).

*Nucleic Acid Based Testing*
While the specific details of NAT are often complicated, the general principle is quite simple and is one that has been implemented since around 1985. The process of testing involves obtaining a nasopharyngeal sample, converting the RNA sample to DNA using reverse transcriptase, and then amplifying the DNA strands using PCR. Understanding the finer details of this process will provide us with insight as to why this form of diagnostic technology is so essential, and how it can also result in false negative tests (Chan et al, 2020).

*Reverse Transcriptase Polymerase Chain Reaction (RT-PCR)*
The process of conducting RT-PCR involves first extracting viral RNA and mixing it with a lysis buffer containing phenol and guanidine isothiocyanate. The combination of these two biomolecules is required to lyse the RNA while also denaturing RNase enzymes. RNase inhibitors

are also included to ensure isolation of viral RNA. This sample is then mixed by pulse-vortexing and incubated. During the incubation period, the RNA is lysed. Then, a purification procedure is completed using a spin column. A spin column is a method used to purify nucleic acids. The spin column is put in a centrifuge, which is a piece of equipment which uses the centrifugal force to separate fluids based on density. The stationary phase of the centrifuged sample is composed of a silica matrix. RNA molecules bind to the silica gel membrane. Proteins and other contaminants are not retained. After centrifugation, the supernatant is combined with a wash buffer. The solution is centrifuged again, which leaves the RNA bound to the silica gel. Once the sample is washed, the column is placed in a clean microcentrifuge tube and an elution buffer is added. The solution is centrifuged, now removing the viral RNA from the spin column. The purified RNA is obtained. It is combined with a master mix that contains a reverse transcriptase. During this reverse transcription phase, the first strand complementary DNA synthesis is primed with the PCR reverse primer, which hybridizes to a complementary region of the RNA genome. Reverse transcriptase then synthesizes the DNA complementary of the viral RNA, now called a cDNA. The RNA strand is removed. The reaction mixture is loaded onto a PCR plate, which is placed into a PCR machine. PCR consists of a series of thermal cycles, with each cycle consisting of a denaturation, annealing, and extension phases. In the denaturation step, the reaction chamber is heated to 95° C. In the next phase, the reaction temperature is lowered, allowing the annealing of the forward primer to its complementary part of the single stranded DNA template. In the extension step, the *taq* DNA polymerase synthesizes a new DNA complementary DNA strand. This cycle keeps repeating until the DNA is appropriately amplified. Following the first cycle, a mechanism is put in place whereby each replication results in the release of fluorophores, allowing the scientist to track how much a certain target DNA is present in real time. This allows for amplification of target sequences in viral RNA (Chan et al, 2020).

An article published in the Journal of Clinical Microbiology highlights the research process involved in developing the most effective RT-PCR diagnostic technology. This study examined the performance of three types of RT-PCR tests that targeted different genes of SARS-CoV-2.

These examinations were facilitated by the early detection of the genomic sequence of this virus. In late January of 2020, the first RT-PCR assays were analyzed targeting the RNA-dependent RNA polymerase (RdRp), envelope (E), and nucleocapsid genes (N). It was found that the assays targeting RdRp, an enzyme that catalyzes the synthesis of RNA, had the highest sensitivity and thus effectiveness. In order to determine whether a patient has tested positive for COVID-19, there must be a positive NAT for at least two of the specified genome targets (RdRp, E or N). The N, or nucleocapsid gene is also a common target, as this gene carries the direction for making the nucleocapsid protein (Chan et al, 2020).

While RT-PCR is a simple method by which the genome of SARS-CoV-2 can be amplified and identified in a sample, the effectiveness of the test has been shown to vary dramatically. It has been shown that in the first few days of the infection, before symptoms even emerge, the chance of getting a false negative on the first day is 100%, and on the fourth day is 67%. On the fifth day of the infection, the probability of getting a false negative is 38%, and 20% on the eighth day. After this point, the chances rise with every day. Thus, there is a one in five chance of getting a false negative test result. According to the WHO, some reasons for false negative COVID-19 tests may include poor quality of specimen containing inadequate amounts of DNA, early or late collection of specimens, inappropriate shipping and handling of specimen, and technical difficulties in running the test (Pan et al, 2020).

*Serological Testing*
The second type of testing method is serological testing, which determines whether antibodies specific to COVID-19 are present. Thus, these tests do not determine if someone is currently infected with the virus, but rather, if they have been infected in the past and produced antibodies. It is performed by taking a blood sample rather than a nasopharyngeal swab. While NAT tests operate by way of RT-PCR, serological tests use another common biotechnology called ELISA, or an enzyme-linked immunosorbent assay. ELISA is a technique used to detect and quantify peptides, protein, antibodies and hormones. The method involves analyzing protein samples that are immobilized in polystyrene plates which passively bind antibodies and proteins. First,

the antigen solution is added to the wells of the polystyrene plates and the plates are incubated. The incubation allows the antigens that hydrophobically adhere to the plastic. The liquid is removed from the wells. Then, irrelevant proteins are added to the wells to cover all the unsaturated binding sites of the microwells. The plate is then incubated with antigen-specific antibodies. The antigen is detected by either a labeled primary or secondary antibody. The current testing kits on the market test for the presence of IgG and IgM in human serum. To explain this, when a person gets infected with any pathogen, the immune system generates two main responses – a primary response and a secondary response. The primary response occurs when the immune system comes in contact with an antigen for the first time and undergoes mass proliferation of B cells, specifically plasma cells, in order to develop antibodies that have a high antigen-binding affinity. The first antibody produced upon exposure is called IgM, and the second type is called IgG. IgG is also the main antibody produced upon antigen re-exposure (Morozov,2020).

One of the prevalent ELISA Antibody Tests consist of two serial direct ELISA's to detect IgG. The first ELISA is performed against the recombinant receptor binding domain of SARS-CoV-2. The results of this are confirmed using another ELISA against the spike protein of the virus. The antibody ELISA test is extremely important in aiding with diagnosis, management and recovery form COVID-19 cases. A positive result for this test shows that the patient has antibodies to COVID-19, which indicates that they have had the virus at some point in time. This result is possible because the IgG antibodies are able to stay in the blood for long periods of time. A negative result means that the patient either does not have COVID-19, or it is too early in the infection to be positive. Some advantages to using ELISA include the fact that it is simple, cheap and well-established. The results can be quickly obtained, and the test can be performed on multiple samples simultaneously. The disadvantages of this test are that there is a higher possibility of obtaining false positives. This is because the antigens being utilized are not well-established and there are several different companies producing these kits at once. Another reason for false positives includes cross-reactivity from pre-existing antibodies. Several biotechnology companies are designing tests that eliminate the variable of cross-reactivity; however, none have been FDA approved thus far (Aydin, 2020).

The research surrounding COVID-19 diagnostic technologies skyrocketed between January and July 2020. Within the first five months of the virus being present, over 930 research papers were published. In comparison, only nine publications were available regarding Zika when it was first discovered. By December 31$^{st}$, 2019, the genomic sequence of the virus was already detected, whereas for the Ebola virus outbreak, it took nearly 3 years. The early development of the genomic sequence of COVID-19 allowed for diagnostic technologies such as RT-PCR to become widely available (Aguiar et al, 2020). Above all the scientific research that has been conducted, the spread of this information through technology during the pandemic has allowed research to progress at a rate that has never been seen before.

# Chapter 3: Ventilators

The demand for advanced healthcare technology has increased exponentially in a short period of time since the onset of the COVID-19 pandemic. Especially in the midst of a novel virus, these tools and devices can be life-saving and shortages will result in devastating outcomes such as increased mortality rates and danger to both patients and healthcare workers. One of these coveted machines, among many others, is the ventilator, which helps provide necessary forms of life support for patients affected by the virus. This chapter will explore the fundamentals of ventilators including its uses and potential risks, its importance in treating COVID-19 patients, and will finally illustrate examples of current innovations and technological advances that are being conducted to overcome the shortage crisis.

## An Introduction to Ventilators

Ventilators are used by medical professionals to either aid or completely take over the work of breathing for patients that are unable to adequately intake oxygen and air on their own. There are two main types of ventilators when treating patients, invasive and non-invasive, both of which are important to understand for the purposes of this chapter. It is also important to note that ventilators are merely supportive therapy, meaning its main purpose is to aid in breathing while healthcare professionals address symptoms with medication and other methods in hopes of helping the patient recover (Manthous & Tobin, 2017).

### Invasive Ventilators

Invasive ventilation is commonly used for patients with more severe cases of COVID-19 and involves inserting a plastic endotracheal tube through the patient's mouth and into the windpipe, also referred to as the trachea. While the patient is connected to a series of monitors, this plastic tube is then connected to the ventilator itself which pushes air and oxygen into the patient's lungs as needed. If required, doctors and nurses are able to remove mucous from the trachea through suction. The ventilator also maintains a steady amount of low pressure, better known as positive end-expiratory pressure (PEEP), to prevent air sacs within the lungs from collapsing (Manthous & Tobin, 2017).

However, invasive ventilators pose additional risks to the patient. First, patients will be more prone to infections such as pneumonia and may be forced to remain on the ventilator for an extended period of time. Patients are susceptible to these infections due to the endotracheal tube which may allow bacteria to enter the body. As well, a section of the lung can also become weak, causing a hole to develop which allows air to leak out of the lung. This results in a collapsed lung, also known as pneumothorax, which must be treated with a tube that is placed into the chest to drain out the leaking air (Manthous & Tobin, 2017).

*Non-Invasive Ventilators*
Other patients that require assistance with their breathing are often treated with non-invasive methods. While there are two types of non-invasive ventilation, this chapter will focus on non-invasive positive-pressure ventilation (NIPPV). Currently, NIPPV methods are contraindicated for COVID-19 patients since these masks emit large volumes of air exhaled by the patient which would contain droplets of the virus (Ranney et al., 2020). However, its benefits to COVID-19 patients will be further explained in a later section of this chapter.

NIPPV involves the delivery of oxygen either variable or constant pressures using a face mask. More commonly used for treatments in sleep apnea, Variable Positive Airway Pressure (VPAP), sometimes called Bi-level Positive Airway Pressure (BiPAP), and Constant Positive Airway Pressure (CPAP) both deliver pressure to help ease the work of breathing for the patient. However, VPAP differs from CPAP since it uses two pressures, a high Inspiratory Positive Airway Pressure (iPAP) and a lower Expiratory Positive Airway Pressure (ePAP), in contrast to CPAP's singular constant pressure (Hormann et al., 1994).

Another method of non-invasive ventilation is the use of a Bag-Valve-Mask (BVM) which involves the repeated action of manually compressing a self-inflating bag to push air into the patient's lungs (Bucher & Cooper, 2018). This method differs from CPAP and BiPAP modes of ventilation in the sense that it is not automated and must be manually performed.

A risk that may be associated with these types of ventilation include intrinsic PEEP, also known as auto-PEEP. As aforementioned, PEEP is the low pressure exerted by the ventilator and is used to provide additional support for the patient's lungs and help alveoli, the air sacs in the lungs, remain open. Auto-PEEP occurs when patients fail to completely exhale the volume of air that was supplied by the ventilator before the next breath is delivered. In this case, a portion of each new breath is retained in the patient's lungs and are therefore at risk of overinflation. The occurrence of auto-PEEP lessens the volume of air the patient is able to draw in on the next breath and the gradual accumulation of pressure in the lungs can be detrimental to the patient's health (Jackson, 2019).

*Ventilators and COVID-19 Patients*
Contracting the novel virus often causes a respiratory infection that makes it difficult for those infected to breathe, making ventilators vital to saving lives. While not all COVID-19 patients will require ventilators, a study conducted in the Seattle region investigating critically ill patients, specifically those of older age (64±18), found that those with more severe cases are often placed on ventilators (Bhatraju et al., 2020). As previously mentioned, ventilators are not considered treatments for illnesses such as the COVID-19 virus; they are used as support to aid the patient's breathing and by extension, other bodily functions, while medical professionals attempt to treat the underlying issues presented by the virus.

However, the challenge that is presented with lung infections associated with COVID-19 is that there are currently no known effective treatments, meaning that patients may be required to use a ventilator for an unknown length of time (Hensley, 2020). Some patients with severe cases may become dependent on ventilators as their illness progresses and become unable to discontinue their ventilator support; others are able to recover and proceed to medical care in the absence of breathing assistance (Manthous & Tobin, 2017). The exponential growth of the pandemic and the increasing demand for ventilation support has resulted in ventilator shortages around the globe (Ranney et al., 2020).

**The Shortage Crisis**

The deficit in ventilators has already proven to exhibit detrimental consequences in regions around the world. The lung infections associated with the virus place ventilators in high demand with no certainty as to when the patient might be able to discontinue its use. Additionally, testing for the virus also directly affects the availability of ventilators and contributes negatively to the shortage. Patients may be required to use ventilators while they await their test results instead of using NIPPV since, as aforementioned, the masks on NIPPV may potentially cause aerosolization of the virus (Ranney et al., 2020). Unfortunately, areas with scarce supplies of ventilators, such as Italy at its peak in the pandemic, are being forced to choose which patients will be admitted to the intensive care unit (ICU), often leaving older patients, or those with other severe health complications, to their own devices. The consequences of ventilator deficits also extends beyond COVID-19 to other illnesses that require critical care as ventilators are now being prioritized for younger patients without other health conditions that have a projected higher chance of survival (Saunders, 2020).

Unfortunately, increasing the supply of ventilators is no simple task and presents itself with many barriers. Firstly, hospital-grade ventilators are expensive to obtain and can cost from anywhere between USD$25,000 and USD$50,000 (Porpora, 2020). This limits the number of ventilators that less developed and less wealthy regions are able to obtain, thereby limiting their capacity within hospitals and making the onset of a pandemic more deadly. As well, another obstacle that was more prevalent towards the beginning of the pandemic was the disruption of the global supply chain. This issue is similarly applicable to other necessities such as personal protective equipment (PPE), however, the inability to transport ventilators has worsened the existing shortage (Ranney et al., 2020). Finally, manufacturers of ventilators are also struggling to keep up with the sudden increase in demand. Countries around the world are demanding thousands of ventilators at a time and manufacturers simply lack the capacity to produce such a volume in a short period of time (Ranney et al., 2020).

**Innovations in Ventilator Technology**

Amidst the global shortage, various institutions and organizations have generated a multitude of innovative ideas to address the issues previously mentioned affecting ventilator supply. For the purposes of this chapter, the innovations concerning altered VPAP machines, robotically pushed BVMs, and finally a shift in production within automotive manufacturers will be explored.

*Modified VPAP Ventilators*
As mentioned earlier, VPAP ventilators are a form of NIPPV that uses two pressures (iPAP and ePAP) to help the patient breathe. This method of ventilation is commonly used in treatments for sleep apnea and does not involve an endotracheal tube. In the standard VPAP machine, the patient does not need to be intubated as a mask is placed on their mouth and nose, however, the mask includes a deliberate air leak. For this reason, the VPAP machine cannot be simply connected to an endotracheal tube and function identical to a standard invasive ventilator; VPAPs are optimized for masks and emit large volumes of contaminated exhaled air.
However, a team of anesthesiologists, pulmonologists, sleep and critical care specialists, and medical students at the Mount Sinai Health System in New York have been able to re-engineer the VPAP machine to serve as a secondary ventilator option. In this project, they have relocated the holes within the circuit and attached a series of filters so that the air being expelled by the patient, and subsequently the machine, no longer contains virus droplets. This altercation ensures that the virus is not being spread as the patient is treated and minimizes the risk towards healthcare workers (Kunzmann, 2020).

This project aims to not only ease the shortage crisis in pandemic hot spots such as New York, but also provide aid for rural and less developed areas that may have limited resources amidst high patient counts. In contrast with the premium price tag associated with invasive ventilators, these VPAP machines will cost around USD $3,000

(Kunzmann, 2020). Many areas might not even require to purchase new VPAP machines; many VPAP machines are already in circulation due to their importance in sleep apnea treatment and just need to be slightly adjusted for hospital use. Simultaneously, the team at Mount Sinai is developing an informational guide so that other hospitals around the globe will be able to assemble their own modified VPAP alternative. This guide will include how to reconfigure the VPAP machine as well as a list of suppliers to turn to for parts. Additionally, the guide explains how to substitute parts that are unavailable in the current disrupted supply chain. Resources such as alternative suppliers and even 3D printing files are made available in hopes that overwhelmed regions will be able to replicate the modified VPAP (Kunzmann, 2020).

It is important to note that the modified VPAP is not able to perform as a perfect substitute for invasive ventilators. Since it is non-invasive, the VPAP is not physically able to move gas in and out of the patient. However, the team hopes that this machine can be used for patients once they show steady signs of recovery, thereby liberating a ventilator for use on another critically ill patient (Kunzmann, 2020).

*Robotically Pushed BVMs*
A team at the Massachusetts Institute of Technology similarly wanted to find an accessible and affordable alternative to the traditional ventilator. The MIT Emergency Ventilator Project was started in light of the primary objective to create an extremely low-cost ventilator. Although, similar to the modified VPAP, the emergency ventilator cannot replace an ICU ventilator, its purpose is to help free up ventilators in life-or-death situations in dire circumstances (MIT Emergency Ventilator, n.d.).

When a hospital has exhausted their supply of ventilators, their last resort is often to assist the patient via manual bagging through the use of a BVM. As aforementioned, BVMs involve the repeated manual compression of a self-inflating bag to push air into the patient's lungs. Generally, BVM ventilation can be performed by one provider, however, a second provider may help squeeze the self-inflating bag as the other holds the mask seal (Bucher & Cooper, 2018). In this case, valuable skilled healthcare workers are occupied with pumping the bag while their assistance may be needed elsewhere.

The emergency ventilator strives to solve this issue by integrating a robotic arm to the BVM. As the patient is connected to a series of monitors, the robotic arm would automatically compress the bag to help the patient breathe and relax when needed. This solution allows patients in a less critical state to be cared for by less specialized healthcare workers while resources are allocated to those who are most in need (MIT Emergency Ventilator, n.d.).

However, the length of time that a patient is able to remain on manual ventilation is unknown. As of now, manual ventilation is simply a short-term solution since there is a lack of clinical evidence surrounding the safety of long-term use (MIT Emergency Ventilator, n.d.). Risks that may arise include the aforementioned auto-PEEP, in which the accumulation of pressure within the patient's lungs can be dangerous. Ultimately, the emergency ventilator's primary purpose is not sustainable long-term care such as the modified VPAP, but instead a cost-effective last resort alternative when all other options have been exhausted.

*Shifts in Production*
As mentioned above, one of the obstacles in the ventilator shortage is the limited capacity that ventilator manufacturers have to produce these machines. To help combat this issue, automotive and transportation manufacturers have begun to shift their production towards helping supply more ventilators to hospitals. For the purposes of this chapter, the efforts of companies specifically located in North America will be discussed.

Concurrent with the production of PPE by renowned clothing brands such as Canada Goose, other industries in Canada are also joining forces with local medical institutions to help in efforts against the virus. Across Canada, companies such as an aerospace and defence supply and training company, CAE, and supplier of auto parts, Linamar, are collaborating with medical suppliers such as Thornhill Medical and StarFish Medical to make an estimated 30,000 ventilators (MacCharles,

2020). Other organizations such as Magna International Inc., Martinrea International, and the Automotive Parts Manufacturers' Association of Ontario agreed to supply 10,000 ventilators for Ontario, one of the Canadian epicentres of the pandemic (Layson, 2020).

Under the Defense Production Act in the United States, the Department of Health and Human Services has a USD $500 million contract with General Motors to produce 30,000 ventilators to aid efforts in treating patients affected by COVID-19. Other automotive companies such as Ford and Tesla have also committed to adding as many as 50,000 ventilators to GM's promise. Ford specifically has explained its plans to make the Airon Model A-E ventilator which requires no electricity and operates solely on air pressure (Massey & Cole, 2020).

**A Future with Ventilators**

The importance of ventilators in the critical care of a patient has never been more emphasized than when patients are in its absence. Amidst the current pandemic, ventilator shortages have become a household term as more and more individuals begin to educate themselves about these complicated machines and the consequences of having too few.

An article written surrounding the effects of the virus on different countries stated that mortality rates are significantly lower in countries whose governments previously invested in large numbers of intensive care beds and ventilators (Saunders, 2020). While the rapid development of projects such as the modified VPAP and the emergency ventilator mentioned above may help save countless lives, the consequences of poor foresight have never been more apparent. That being said, the future may involve more research, development, and funding towards intensive care resources.

Moving forward, this might include more projects dedicated to innovations and advancements surrounding technology used in critical care. More manufacturers and supply chains might also be established to increase the production of ventilators, PPE, and other medical necessities. These supplies are not only important in containing a pandemic but are also necessary to treat other critical illnesses on the rise. Additionally, while the mass production of ventilators are not able to serve those that were struggling at the pandemic's peak, its circulation, especially in rural and less developed areas, will provide support in the event of an influx of patients from an anticipated second wave. Projects such as the modified VPAP and the emergency ventilator also use existing devices and equipment that can easily be repurposed, making them versatile for use in both COVID-19 and other intensive care patients. Ultimately, the pandemic's impacts have undoubtedly demonstrated the importance of life-saving ventilators and the need to proactively invest in critical care technology.

# Chapter 4: The Role of Masks

The onset of COVID-19 has fundamentally changed how society views, uses and produces masks. They are a critical tool in protecting healthcare workers and stopping the spread of the pandemic by reducing the overall likelihood of transmission of the virus between individuals. Unfortunately, social distancing on its own is not one hundred percent effective, as out of 10 studies measuring horizontal droplet distance, 8 showed droplets travelling more than 2 meters with some showing travel as far as 8 meters. Additionally, droplets containing the virus in a viable form were found in the air up to 3 hours after they were initially expelled (Bahl et al., 2020). This highlights the desperate need for masks not only for healthcare workers but also for the general public in order to reduce the spread of the virus on a population level, as COVID-19 is a respiratory illness that primarily relies on aerosol transmission to spread. Unfortunately, there are many cases of individuals being asymptomatic carriers, meaning that they are infected and able to spread the virus even though they do not show the symptoms. As a result, it is important that everyone wears a mask if they are able to, regardless of whether or not they are symptomatic. This chapter will highlight the need for masks, how different types of masks and respirators work, and the changes that have been made to mask production as a result of the pandemic along with its potential consequences.

## The Types of Masks and How They Work

The general purpose of masks is to reduce the spread of aerosol-transmitted pathogens, such as COVID-19, by covering the mouth and nose of an individual to reduce the likelihood of the wearer either contracting the virus or spreading the virus to someone else. There are 3 main types of masks: N95 respirators, surgical masks, and cloth masks. According to a 2008 study, the use of masks, regardless of type, all reduced aerosol transmission. However, each of these masks have varying degrees of outward protection, which is the extent to which they prevent the wearer from spreading the virus to the surrounding environment, and inward protection, which is the extent to which they protect the wearer from contracting the virus if it is present in the surrounding environment (van der Sande et al., 2008). The effectiveness of a mask with regards to reducing the transmission of a virus can primarily be attributed to 3 factors: the transmission-blocking potential of the material used, the fit and related air leakage of the mask, and the degree of adherence to proper wearing and disposal of masks (van der Sande et al., 2008). Each of the 3 types of masks have their own advantages and disadvantages and are best suited for different situations.

N95 respirators have proven to be the most effective at providing a barrier to aerosol transmission and are the only masks that provide effective inward and outward protection. They are able to filter 95% of small particles, including droplets containing COVID-19, out of the air (Mayo Clinic, 2020), meaning that the transmission-blocking potential of the materials used is high. In order to be effective, however, the mask must have a tight seal around the wearer's face to prevent air leakages (FDA, 2020). These masks also have their disadvantages, as they are currently in short supply and may become difficult to wear for prolonged periods of time due to the airtight seal (Seladi-Schulman and Weatherspoon, 2020). It is for these reasons that governments are recommending that the general public does not wear N95 respirators and that they are instead directed towards healthcare workers. Not only do members of the general public not have adequate training to properly fit and seal their mask, which is crucial to its ability to block the transmission of COVID-19, but they are also far more likely to encounter factors and situations in their everyday lives - such as going out to restaurants to eat - that interfere with the proper wearing of an N95 respirator (van der Sande et al., 2008).

As expected, N95 respirators differ greatly in their construction when compared to surgical and cloth masks. They consist of 2 outward layers of protective fabric, a pre-filtration layer, and a highly efficient melt-blown filtration layer. The layers of protective fabric, which act as protection against the outside environment and as a barrier to anything in the wearer's exhalations, are made through a process known as spun bonding. This process uses nozzles to blow 15-35 micrometre polymer threads onto a conveyor belt, which form a cloth as the belt continues before being thermally, chemically, or mechanically bonded. The pre-filtration layer is usually a needled non-woven layer, meaning that barbed needles are repeatedly sent through the non-woven material to hook fibres together and increase their cohesiveness. The material is then calendered, which is the process of compressing and smoothing materials by running a single continuous sheet through multiple pairs of heated rolls. This process thermally bonds the plastic fibres, making the layer stiffer and thus easier to mould. The final layer is a highly efficient melt-blown polymer layer which is primarily responsible for filtering out small particles. The melt-blown process is similar to the spun bonding process, with the primary difference being that the polymer fibres leaving the nozzles are less than a micron wide. Although the melt-blown fabric is sometimes thermally bonded to add strength and abrasion resistance, the fibres generally bond on their own as they cool on the conveyor belt. This layer is also usually treated to be an electret, meaning that electrostatic properties are added to the filter layer, allowing for electrostatic adsorption to trap aerosolized particles via electrostatic attraction. The final assembly occurs through the use of converting machinery, which combines the layers using ultrasonic welding and adds straps and metal strips to the mask (Henneberry, n.d.).

*Surgical Masks*
Surgical masks protect the wearer from splashes and large-particle droplets while preventing the transmission of potentially infectious respiratory secretions from the wearer to others. However, they are not effective at protecting the wearer from infections such as COVID-19 due to the large filter size and gaps on either side of the mask which result in air leakages during inhalation (Seladi-Schulman and Weatherspoon, 2020), meaning that although they provide a relatively high degree of outward protection, they provide little inward protection. Despite this, they are still recommended for routine medical procedures, as the World Health Organization (WHO) stated that "a medical mask is recommended for routine care, whereas a respirator (airborne precautions) is recommended if [healthcare workers] are conducting an aerosol-generating procedure such as endotracheal intubation, bronchoscopy or airway suctioning, along with droplet precautions" (WHO, 2020b). Consequently, surgical masks should also not be worn

by the general public, as they are required by healthcare workers and in high demand by hospitals.

Although they offer significantly less protection to the wearer, surgical masks share some similarities to N95 respirators with regards to the materials used for construction as they consist of 3 non-woven layers stacked together. The inner layer, which comes into contact with the wearer's face, is designed to trap and moisture that is expelled from the wearer's exhalations whereas the outer layer serves as a waterproof barrier that prevents any liquids expelled by others while talking, coughing, or sneezing from being transmitted through the mask. The center layer is responsible for air filtration, and much like N95 respirators is made from melt-blown polymers that are treated to be an electret (Edwards, n.d.).

*Cloth Masks*

Although cloth masks may be the least effective with regards to blocking aerosol transmission, they play a critical role in the fight against the spread of COVID-19. This is because while other masks are in short supply, cloth masks are readily accessible and can either be purchased from a variety of retailers or home-made. The main purpose of cloth masks is to prevent droplets expelled by the wearer from reaching the surrounding environment, meaning that they provide some degree of outward protection and can reduce the spread of COVID-19 when widely used by the public as many individuals may be infected yet asymptomatic. However, it is important to recognize that cloth masks provide little to no inward protection and that their purpose is solely to prevent the wearer from spreading the virus, not to protect them (Mayo Clinic, 2020). Currently, the Center for Disease Control (CDC) recommends that all members of the general public wear cloth face masks in public regardless of whether or not they feel sick. This idea is now commonplace in many parts of the world, as numerous countries and regions have even mandated the wearing of face coverings in public. Cloth masks are being recommended not only because of their abundance, as there are now even tutorials online for how to make one from common items, but also because of the fact that surgical masks and N95 respirators must be reserved for healthcare workers and frontline staff (Mayo Clinic, 2020).

**Mask Production**

The COVID-19 pandemic has brought about a sharp increase in the demand for all types of masks and personal protective equipment (PPE). Unfortunately, most types of PPE, including masks, are heat sensitive and thus must be single-use only as they cannot be reprocessed using sterilization processes that are commonplace in many hospitals as they typically use high levels of heat (Rowan and Laffey, 2020). As COVID-19 began to spread throughout the world in early 2020, many hospitals and organizations were left scrambling to meet equipment demands, leaving some practitioners with no choice but to reuse what would normally be disposable masks for a prolonged period of time. Thankfully, the situation has since then improved, but mask production is by no means at the ideal level even several months into the pandemic. Companies and governments alike have had to make dramatic changes to both production and policies, and although there has been a great improvement in the availability of masks and other PPE, there is still a long way to go.

*Measures Being Taken to Increase Production*
The most obvious and arguably effective approach for companies has been to substantially increase production. In fact, the WHO estimated that the global demand for masks was at around 89 million per month back in March 2020 and that companies would have to increase production by a staggering 40% to meet this need. This shortage has resulted in the price of surgical masks increasing sixfold and the price of N95 respirators increasing threefold (WHO, 2020c). In the United States alone, the demand for N95 respirators stood at approximately 50 million per year, but this number has since skyrocketed to 140 million per year according to the Department of Defense. The result of this is a substantial increase in the production of masks, as the United States is expecting to reach 450 million masks per year by October and 800 million per year by January 2021 (Lopez, 2020).

There are many more measures beyond simply increasing production that can be taken to increase the availability of masks to healthcare workers. The first measure is to encourage and facilitate the donation of PPE such as masks to healthcare facilities, as there are numerous industries such as construction that may have stockpiles (Ranney et al., 2020). The hope with this is to maximize the use of existing stockpiles to reduce the burden on the production chain, as these other industries do not have as high of a need for PPE during this time. Another measure that has been taken by the United States Government, and may influence other nations, is to reduce the Food and Drug Administration (FDA) standards for N95 respirators, as many of the millions of masks being produced are never brought into circulation due to not meeting FDA standards (Ranney et al., 2020). Although this measure does increase the availability of N95 respirators to healthcare workers without needing to directly increase production, it is not without its disadvantages. The purpose of the FDA is to ensure that products are of high quality such that they are able to effectively protect the wearer. With a reduction of these standards, there is the risk that many sub-par N95 respirators could now be distributed, and although this idea behind this measure is that giving practitioners something is much better than nothing at all, it may put more healthcare workers at risk of contracting the virus.

The increased demand for masks as a result of this pandemic has also resulted in substantial innovation, as many companies are now experimenting with the use of 3D printing technology to produce masks and increase their availability to frontline workers. The main advantage of 3D printing beyond simply increasing production capacity is that they can be custom made for an individual through the use of 3D laser scanning to scan exact facial parameters. The result of this is a mask that is not only more comfortable but also provides a better seal (Ishack and Lipner, 2020). Looking at a specific example, a team has designed a partially reusable N95 mask for use by healthcare workers. The mask consists of 2 reusable polyamide composite components - the facemask and filter membrane support - along with 2 disposable components, which are the filter membrane and head fixation band (Swennen et al., 2020). By reducing the disposable components to only those which are easily mass-produced, the production of this mask will hopefully be far more efficient in comparison to traditional N95 Respirators. This is because all materials needed to produce this mask are readily available. The 3D modelling of a person's face can be done by designers world-wide using free, downloadable CAD (Computer-Aided Design) software, the disposable melt-blown filters are globally available from manufacturers already producing FFP2/3 protective masks, and the velcro straps used for the headband are widely produced and commercially available. However, leakage and virological tests of this prototype following one or more disinfection cycles have not yet been performed, and they are a critical step before the mask can enter production (Swennen et al., 2020). Regardless, this mask is an excellent example of how innovation and the repurposing of existing technology can be used to aid the fight against a global crisis.

*Environmental Consequences*

Although a substantial increase in the production of single-use masks is necessary, there are now growing concerns regarding how this will affect our environment. Unfortunately, single-use polymeric materials, such as those used in the production of masks, are a significant source of plastic and plastic particle pollution in the environment, with an estimated 60-95% of all ocean litter being plastic (Schnurr et al., 2018). This means that masks may be a major source of microplastic pollution, as they can degrade or break down into particles smaller than 5 mm, which are known as microplastics under environmental conditions (Fadare and Okoffo, 2020). The plastic particles introduced by the disposal of masks can then go on to have a multitude of severe consequences on the environment and surrounding ecosystems. In aquatic ecosystems, plastic pollution can have disastrous effects, with marine fauna being at risk of either becoming entangled in the plastic pollutants, which has proven to be the most lethal consequence, or ingesting them. Additionally, there is the risk of chemical contamination, which although proven to be significantly less lethal, still has significant damaging effects (Wilcox et al., 2016). Overall, it has become evident that the widespread use and mass production of masks is necessary, but a greater focus must be placed on the safe disposal of masks to prevent them from filling up landfills and harming both terrestrial and aquatic ecosystems.

As the COVID-19 pandemic continues to dominate the world stage, one of the most simple and basic medical tools available to us has proven to be one of the most valuable. It has not only allowed healthcare workers and first responders to treat the infected with a reduced risk of contracting the virus themselves, but it has also greatly assisted with efforts to contain the pandemic and reduce its spread by promoting, and in some areas mandating, the use of masks by the general population. Although this pandemic has wreaked havoc on millions and killed hundreds of thousands, the situation could have been unimaginably worse without the availability of face masks.

# Chapter 5: Medical Technologies and Treatments

Thus far, we have established an understanding of the basic diagnostic technologies and equipment used in the prevention of COVID-19. Serological and nucleic acid-based tests allow us to determine whether an individual is currently infected with the virus or if they have been infected in the past. Masks have been shown to be important in helping us prevent others from acquiring the virus, and ventilators are essential in maintaining the health of COVID-19 patients. However, an area of research that is causing a great deal of stress among research worldwide is the treatment of COVID-19. It has been argued that the end of this pandemic will be marked by the development of a vaccine that must be made mandatory for the entire population (Callaway et al, 2020). At the same time, there have been several speculative treatment options that made it on the market such as Remdesivir and Hydroxychloroquine. Analogous to the battle between the US and Russia for the mission to the moon, the race for a treatment or vaccine for this virus will be one celebrated for centuries to come.

## Types of Vaccine Technologies

The development of a vaccination for COVID-19 has been fast-tracked due to the urgency of the situation, as well as the medical technologies available. We will begin our examination of medical technologies in the pandemic with an overview of vaccinations in general, and a deeper look into which types of vaccines companies around the world are investigating.

### What is a Vaccine?
In simple terms, a vaccine is a product that generates an immune response in order for a person to develop immunity to a specific disease. Immunity is developed when the body's adaptive immune system recognizes new pathogens and develops antibodies and memory cells

that will fight the pathogen upon re-exposure. Before we look at the different vaccines being developed, we will examine how SARS-CoV2 immunity may be developed. After the virus enters the body, it uses a protein on its surface called the spike protein, which binds to the ACE2 receptor on the surface of human cells. This receptor essentially acts as a gateway for the virus's genetic material to enter the human cell. When the virus is in the cell, it hijacks the cell and causes it to translate its RNA into proteins. The virus then reassembles and is released from the cell. Then, as a part of the body's immune response, antigen-presenting cells, also called APCs, phagocytose the virus and present it to T-helper cells. The T-helper cells then present the antigen to B cells, which are able to produce antibodies. These antibodies can either directly lyse the antigen or tag the antigen for destruction. Additionally, cytotoxic T cells are able to directly destroy infected cells. This constitutes the primary immune response. As the B cells are activated, long-lived 'memory' B and T cells are also produced that can recognize the virus for varied amounts of time. The goal of a vaccination is to introduce a particular antigen into the body that will not infect the person, but will still provoke an immune response (Callaway et al, 2020) .

*Live Virus Vaccines VS Inactivated Virus Vaccines*
There are currently approximately eight different classes of vaccines being developed. One of the more traditional types of vaccine is the whole virus vaccine. These vaccines require the virus to be grown in large quantities in eggs, and then can be inactivated or live and attenuated. Inactivated virus vaccines utilize formaldehyde or heat/radiation to kill the virus. This type of vaccination has the benefit of being usable on individuals with weakened immune systems. However, it does not generate as strong of an immune response as a live virus. One of the leading vaccine candidates, produced by Sinovac, a leading biotechnology company based in China, is developing an inactivated virus vaccine. This product has made it to phase 3 clinical testing and is a promising candidate. The phase 2 clinical trials showed that individuals were able to produce neutralizing antibodies 14 days after vaccination, and over 90% of 600 volunteers showed an immune response (Liu et al, 2020). On the other hand, the live, attenuated viruses are grown in cells but are weakened instead of killed. This can be done by passing the virus through animal or human cells until it develops

mutations. The live virus vaccine candidates for COVID-19 thus far have utilized an approach called codon deoptimization (Smith et al, 2020). This technique involves rebuilding the virus from scratch and then incorporating certain mutations that weaken it. This type of vaccination is able to provide long-lasting protection after one dose since it is similar to the real infection;however, it may not be suitable for people with other health conditions. These vaccines are also difficult to transport and must be refrigerated, making it inaccessible to countries with limited access to appropriate transportation and storage infrastructure.

## Viral Vector Vaccines

Another class of vaccines being researched are viral-vector vaccines. There are two types of viral-vector vaccines - those that can replicate within cells and those that cannot. These vaccines use the virus as a vector to transport DNA into human cells. This DNA encodes the antigen, which is able to elicit an immune response. This immune response includes antibody, CD4+ T cell, and cytotoxic T lymphocyte-mediated immunity. The viruses are genetically modified to eliminate pathogenicity. They are also engineered to carry a certain protein from the disease being eliminated, such as the spike protein from SARS-CoV-2. This biological machinery was first developed in the 1970s to infect monkey kidney cells. The Ebola vaccine is an example of a viral-vector vaccine that is able to replicate within cells. This vaccine is able to provoke a very strong immune response. The virus can be modified to have certain mutations that prevent it from replicating, but these vaccines do not produce long-lasting immunity and several booster shots may be necessary. Additionally, individuals who have already been exposed to the viral vector may be resistant (Ura et al., 2020). In general, viral vector vaccines are good at inducing a strong T cell response.

## Nucleic Acid Vaccines

Nucleic acid vaccines are another class of vaccines that are being developed against SARS-CoV-2. These vaccines utilize genetic instructions for the spike protein to prompt an immune response (Callaway, 2020). Hence, the actual antigen is not introduced to the body, only the genetic information is. RNA vaccines inject the instructions to make the spike protein, leading the body to make the protein so that we can generate an immune response. The RNA is often encapsulated with a lipid coat, making it lipophilic allowing for easier entry into cells. RNA vaccines induce a primary immune response, which activates B cells allowing memory cells to be produced. Since the sequence of the SARS-CoV-2 genome has already been determined, RNA can be produced *in vitro* instead of using chicken or mammalian eggs. The production time for RNA vaccines is very low, making them a promising vaccine candidate. However, they do not produce as strong of an immune response (6). DNA vaccines are also being researched. This mechanism involves the introduction of a plasmid, which is an extrachromosomal DNA molecule, containing the DNA sequence encoding the antigen into appropriate tissues. This vaccine stimulated B and T cell responses. A novel technology in DNA vaccines is the use of adjuncts that assist DNA in entering cells and allowing it to target towards specific cells (7). Additionally, a process known as electroporation creates pores in cell membranes, which increases the uptake of DNA into a cell. Once the DNA enters the cells, it is transcribed into mRNA, which is translated into the spike peptide generating the immune response (Eckert et al,, 2020).

*Protein-Based Vaccines*
Next, protein-based vaccines directly inject the spike protein into the body. This technique may also use fragments of proteins or protein shells. Protein subunit vaccines are developed using recombinant genetic technology. One of the vaccine candidates includes a recombinant spike protein. This company plans to insert the spike gene into a baculovirus, which infects insect cells. The virus will infect lab-grown insect cells and mass produce the spike protein that is folded as if it were produced by humans. This spike protein will then be inserted into a micelle, which is an artificial membrane. The micelle will display certain epitopes of the protein, allowing for the vaccine to target multiple strains of coronavirus. The vaccine also consists of a Matrix-M saponin-based adjuvant. This adjuvant allows for enhanced immunogenicity and humoral and cellular immunity. It consists of saponins, cholesterol and phospholipids. This matrix induces the migration of APCs to the immunization injection site (Draper et al., 2020). The development of this vaccine is truly remarkable and was able to reach clinical trials in Spring 2020. This was possible mainly due to the vast array of products

produced by Genscript, one of the leading biotech companies in the fight against COVID-19. Genscript synthesized the spike gene in 3 days. The University of Queensland has also partnered with GSK to develop another protein subunit vaccine utilizing a protein molecular clamp to stabilize the spike protein. This involves using an antigenic protein that clamps onto virus proteins to trigger an immune response (Draper et al., 2014). Protein subunit vaccines have the benefit of rapid production, however, they do not generate as strong of an immune response since they require adjuvants. The other type of protein based vaccine is the virus-like particle (VLP). VLPs utilize empty virus shells to mimic the structure of the coronavirus, but lack genetic material. Currently, Medicago, another leading biopharmaceutical company, is designing a VLP technology. While live attenuated virus vaccines use chicken eggs as the production platform for the virus, Medicago has worked towards utilizing plants as a 'protein factory' (Liu et al., 2020). The process of developing this vaccine involves first synthesizing genes from the viral genomic sequence, and then infiltrating the genetic material into plants. The plants are then incubated in growth chambers to allow for VLP formation, and then the plants are harvested to extract VLPs. These VLPs are then able to present antigens to a person's immune system in an efficient manner, eliciting a long-lasting immune response. The drawback of this type of vaccine is that they can be hard to produce in large quantities (Callaway et al., 2020).

As of July 2020, there are 4 vaccine candidates that have made it to phase 3 clinical testing, a process that normally takes more than 10 years. However, it is equally important to search for a COVID-19 treatment, a process which may also take upwards of 10 years. For this reason, the drugs being researched are ones that already exist. Most of these drugs are antivirals, a class of medication to treat viral infections.

**Drug Development**

Remdesivir, a drug developed about a decade ago, has been shown to be safe in humans and block the virus from replicating. It was manufactured by Gilead Sciences, and studies showed that individuals taking the drug recovered from COVID-19 in 11 days. On the other hand, individuals who were not on Remdesivir took 15 days to recover. It was later shown that participants who took Remdesivir showed no benefits. Phase 3 clinical trials will be conducted in an ongoing manner. This is the case for most antivirals being used to treat COVID-19, such as Arbidol, EIDD-2801, Favipiravir, Kaletra and several others. Kaletra in particular is an interesting drug as it combines Iopinavir and ritonavir, two drugs that work against HIV. Studies have shown that this drug combined with ribavirin and interferon beta-1b, clears the virus at a faster rate than participants on a placebo. It is evident that more research must be conducted regarding treatments for COVID-19, however, there are several promising candidates (Radcliffe, 2020).

# Chapter 6: Impact of Artificial Intelligence on the fight against COVID-19

## Introduction to Artificial Intelligence

Artificial Intelligence (AI) is a wide-ranging branch of computer science concerned with building smart machines capable of performing tasks that typically require human intelligence, such as speech recognition, decision-making, and translation between languages (Daley, 2019). While it may seem as though AI is technology of the future, it already exists all around us in many of the applications we interact with on a daily basis, including smart assistants, google search, and spam filters on email servers (Daley, 2019).

## Types of Artificial Intelligence

There are generally 2 broad categories of AI;
> 1.      **Narrow AI:** This kind of Artificial Intelligence operates within limited parameters and is focused on performing a single task extremely well. While these machines may seem intelligent, they operate under far more constraints than even the most basic human intelligence (Daley, 2019).
> 2.      **Artificial General Intelligence/AGI:** This is the type of artificial intelligence we often see in the movies, like the robots from *Westworld*, a popular television show, which have general intelligence and are able to apply that intelligence to solve any problem (Daley, 2019).

Many of the breakthroughs in Artificial Intelligence have come from the use of machine learning and deep learning. Machine learning is a system that "focuses on the development of computer programs that can access data and use it to learn for themselves"(Expertsystem, 2017). A computer is fed data to which it applies statistical techniques to help it "learn" how to get progressively better at a task without having been specifically programmed for that task, eliminating the need for millions of lines of written code (Daley, 2019). Alternatively, deep learning is a type of machine learning that runs inputs through artificial neural networks inspired by the structure and function of the brain (Hinton, Osindero, & Teh, 2006). The computer is taught to filter inputs through different layers that predict and classify information (Bonner, 2019).

## How AI is being used in the fight against COVID-19

*Usage in Diagnosing COVID-19*
One usage of AI in the fight against COVID-19 is in the diagnosis of the infection. Researchers across the world are finding ways to work alongside the technology to drastically change the speed and accuracy of COVID-19 detection in patients. One example of this is at the Lawson Health Research Institute in London, Ontario, where researchers are investigating if deep learning can be trained to learn and recognize patterns in ultrasound lung scans of patients with confirmed COVID-19 by comparing them to ultrasound scans of patients with other types of lung infections (DeLaet & Dubeau, 2020). This task is especially suited for AI because many of these patterns and differences exist in details at the pixel level that cannot be perceived by the human eye (DeLaet & Dubeau, 2020). Another group looking into the technology is a team at Mount Sinai Hospital. They have developed an algorithm that rapidly detects COVID-19 based on CT scans of the chest as well as patient information, including symptoms, age, bloodwork and possible contact with someone infected with the virus (Tkachenko, 2020). The study involved scans of more than 900 patients from institutional collaborators at hospitals in China and it statistically had a significantly higher sensitivity (84%) compared to radiologists (75%) evaluating the data. The Mount Sinai research team is now focused on further developing the model to find clues about how well patients will do based on subtleties in their CT data and clinical information (Tkachenko, 2020).

One problem that many researchers investigating this issue face are imperfections in many of the images fed to machines (Levin, 2020). While humans can easily ignore slight imperfections in an image, these imperfections can confuse even the best machine learning algorithms. These result for a wide variety of reasons, ranging from the placement of black rectangles blocking out patients' personal information to technicians underexposing the X-ray, making the X-ray cloudy (Levin, 2020). However, researchers are working to overcome this issue by enhancing and pre-processing many of the images to tidy the imperfections (Levin, 2020).

Another complication that researchers in this field are facing is that, while AI can identify images that look like COVID-19 and other types of coronaviruses, it cannot indicate why these images meet the criteria from a medical point of view (Levin, 2020). To overcome this problem, researchers are teaming up with radiologists to add medical context to each image.

*Usage in Contact Tracing*
Contact tracing is "a process that is used to identify, educate and monitor" individuals who have encountered someone infected with a particular virus, in this case COVID-19 (Public Health Ontario, 2020). This is a crucial process in the fight against a disease because it provides the best hope for breaking "chains of transmission" that occur as a virus spreads (Thompson, 2020). Artificial Intelligence can make contact tracing more efficient and enable its mass deployment across large populations.

One company looking at using this technology is Facdrive Health, based in Toronto, in partnership with two professors from the University of Waterloo (University of Waterloo, 2020). The platform aims to quickly alert those who may have come in contact with a COVID-19 patient and educate them on the subsequent steps they need to take to get tested. The company also provides the option for an AI-enabled bluetooth device which measures vital signs and monitors the recovery of patients in real time (University of Waterloo, 2020).

Another company attempting to bring this to market is Volan Technologies (Mckay, 2020). The company had previously created an "AI-based location positioning system" for use in schools and corporate offices for emergency situations. It works by creating a mesh network in a building or school that tracks where students and employees are and who they are in contact with. They are now attempting to adapt this to function as a contact-tracing tool which could track a positive COVID-19 patient's movements over the last few days to determine who they had been in contact with (direct tracing) and which rooms they were in that could have been contaminated by someone else (indirect tracing). By using a bluetooth mesh system, this network is more secure in comparison to the pre-existing contact tracing tools that depend on a Wi-Fi network, helping safeguard individuals' privacy (Mckay, 2020).

Another tool that is helping fight COVID-19 on a global scale is EpiRisk (Waltz, 2020). This platform was developed by computer scientists at Northeastern University in Boston, and it estimates the probability that infected individuals in an area will spread the virus to other parts of the world through travel. It has helped track the effectiveness of lockdown procedures and travel bans as well as indicated the countries and regions most at risk for seeing imported cases of COVID-19 (Garrity, 2020).

*Usage in developing treatments*
Artificial Intelligence is also playing a crucial role in the search to find effective treatments and vaccines against the COVID-19 virus. One of the biggest players in this market is Benevolent AI, a company working to integrate Artificial Intelligence into scientific innovation (Simonite, 2020). The company's AI algorithm combines drug industry data with information from scientific research papers and, during the pandemic, it has worked to determine whether pre-existing drugs used to treat other diseases can be used to treat COVID-19. A breakthrough for the company occured in late January, when it was able to identify a drug used for rheumatoid arthritis drug that can lessen a few of the most severe effects of COVID-19. This has led to the initial marketer of the drug, Eli Lilly, to begin work on a large clinical trial in July (Simonite, 2020). Another example of this is taking place at the University of Toronto, where students are similarly using machine learning to determine whether existing drugs can be repurposed to fight COVID-19 (Fraumeni, 2020). To date, they have discovered molecules that can impact the virus's ability to enter cells as well as treatments that can limit the respiratory distress that are experienced by patients of the disease (Fraumeni, 2020).

While the ultimate goal is to develop treatments to target the virus, researchers around the world need a clear picture of the virus's structure. To aid this, AI is also being used to map the virus's structure and identify all of its components. One organization making strides in this area is DeepMind, an affiliate company of Google. Its researchers have used its AlphaFold AI system to predict the structure of "several under-studied proteins" that form SARS-CoV-2, the virus that causes COVID-19 (Jumper, Tunyasuvanakool, Kohli, & Hassabis, 2020).

A problem that all companies attempting to use AI platforms to map the compound's structure or determine potential treatments face is the availability of data to "train" the algorithms (Singh, 2020). Much of the data belongs to individual organizations and sharing is not commonplace. This leads to small pools of data that are not sufficient for artificial intelligence algorithms. However, with increased protocols by governments to foster sharing of data and an increase in the amount of data collected over time, this problem will become less prominent (Singh, 2020).

*Usage in treatment procedures*
Artificial Intelligence is also being used in treatment procedures in countries around the world to assist in patient treatment and recovery. One prominent example of this is being executed by a team out of Boston called Partners HealthCare, which has built an AI-based online screening tool that differentiates between those who might have COVID-19 and those who appear to have other, less threatening illnesses, directing each to appropriate help (Wittbold, Carroll, Iansiti, Zhang, & Landman, 2020). This tool served more than 40,000 patients in its first week and helped decrease the strain on healthcare providers.

Another example is taking place in China, where the technology company, Baidu, has developed a no-contact infrared sensor system that singles out individuals with a fever, even in large crowds, and prompts these individuals to seek further help (Wittbold et al., 2020). The Qinghe railway station in Beijing has adopted this system to identify contagious individuals, helping replace a manual screening process. A similar setup has been created at Florida's Tampa General Hospital, which, in collaboration with Care.ai, has deployed an AI system at all of its entrances (Wittbold et al., 2020). This system uses cameras to conduct a facial thermal scan of visitors and pick up on other symptoms like sweat and discoloration to prevent visitors with COVID-19 symptoms from visiting patients.

China's Wuhan Wuchang Hospital has also established a smart field hospital manned largely by robots. Patient's vital signs are monitored by thermometers and bracelet-like devices while the robots deliver food and medicine to patients. This process helps reduce physician exposure to the virus and reduce the workload of health care workers (Wittbold et al., 2020).

*Usage to fight misinformation during COVID-19*
COVID-19 is the first global pandemic during the technology era, with many people around the world relying on social media for critical information about treatment and prevention. This has also led to a rise in factually incorrect posts that have quickly spread through platforms like Facebook, Twitter and Instagram and which have had serious consequences for viewers who chose to act upon this information (Perry, 2020). In fact, in the month of April alone, fifty million posts about COVID-19 were disseminated on Facebook, with two and a half million of these being ads for face masks, testing kits and other products that were banned from advertising on the platform (Perry, 2020). This has led to an expansion of Facebook's fact-checking department to include human fact-checkers from over sixty fact-checking organizations around the world, as well as the implementation of AI based-systems to detect misinformation (Perry, 2020). These systems drastically reduced the workload of human workers and helped effectively flag misleading posts on the platform. Similar procedures are being implemented on platforms like Twitter to help fight the rise in misinformation during these times.

One problem that all AI-based systems built to fight misinformation experience is difficulty detecting images that appear alike to a person but not alike to a computer (Perry, 2020). For example, if an image was screenshotted from an existing post, the image would appear identical to a human. However, the pixels would be very different to a computer and it would thus have difficulty flagging the post. This has led to many scammers changing a small number of pixels on posts spreading misinformation and then re-sharing them through the platforms (Perry, 2020). To fix this, companies are looking to incorporate "multimodal content analysis tools" which consider both text and images together to interpret a post. Furthermore, AI algorithms are being trained to identify objects in images that violate companies' policies and flag posts with these objects (Perry, 2020). However, companies need to be proactive and constantly adapt their strategies to help control this issue.

# Chapter 7: The Rise of Telecommunication

Since its genesis, telecommunication has grown to become an essential component in the everyday lives of many. As social creatures, it is key for humans to remain connected with each other through different means of communication. Technological advances have made keeping in touch more convenient and are especially necessary when face-to-face meetings are not possible. Especially in the midst of a pandemic, society is in desperate need of salvaging these connections for comfort, support and productivity as technology and its role in the pandemic constantly evolves to fulfill the requirements of interconnected communities. This chapter will explore the basics of telecommunication, the effects of the COVID-19 pandemic on telecommunication companies as well as the general public, and finally extrapolate into the future of the industry.

## An Introduction to Telecommunication

As aforementioned, telecommunication can be used to uphold connections between two or more parties when in-person conversations are not possible. Quite simply, it is the convenient exchange of information through technology that allows for communication over a distance greater than what is feasible with one's voice. Texting, calling, or video chatting are all types of telecommunication which may help individuals keep in touch with loved ones, send updates to their workplace, chat with a friend, and anything in between.

During the COVID-19 pandemic, telecommunication has been on the rise in response to the enforced social distancing regulations since prolonged isolation can have adverse effects on one's mental health. While humans are motivated through their desire for interpersonal attachments and the need to belong, this sense of interconnectedness and belonging is absent in isolation (Baumeister & Leary, 1995). A study in California had compared the levels of bonding that participants felt

43

during in-person interactions, texting conversations, audio and video calls. While in-person communication resulted in the highest level of bonding, other forms of telecommunication still offered an adequate amount of bonding. The study had also discovered that many individuals were also more open and comfortable when telecommuting with someone new; bonding levels with strangers through telecommunication were actually higher than those experienced during in-person interactions (Sherman et al., 2013). Especially during the pandemic, these findings have significant implications towards the importance of telecommunication in these times.

**The Effects on Telecommunication Companies**

Telecommunication companies are one of the largest stakeholders in the rise of telecommunication during the current pandemic. While many of these companies will benefit from the increase in use of their systems, the shift to remote work by many will drive the demand for even more networking services and infrastructure. This heightened demand at a rapidly growing rate may potentially strain the system; telecommunication companies must be able to adapt to this ever-changing environment. While a more thorough economic analysis of telecommunication and technology companies will be discussed in a later chapter, this chapter will primarily be focusing on the immediate changes in the telecommunication industry as caused by the pandemic.

With the rise of telecommunication during the COVID-19 pandemic, it is no surprise that telecommunication companies are experiencing large spikes in their network usage. In these times, network reliability and resilience is a concern for many companies. Europe's network infrastructure has recently experienced a decrease in connection rates as well as lower audio and video quality during calls in efforts to lessen connection drops and outages. Additionally, to boost network resiliency, different telecommunication companies in the United States have partnered together to increase capacity. This unprecedented collaboration between competitors involves the sharing of spectrum across the country to ensure the continuation of network connectivity for its users. Moreover, the spikes in network usage undoubtedly brought with it an influx of revenue, but certain telecommunication companies may actually experience trouble with cash flow in the long term. Many telecommunication companies own sports-related media, which may witness a negative impact on profits driven by advertisements during the current climate of sports league cancellations (Casey & Wigginton, 2020).

However, telecommunications companies are also joining the fight against the COVID-19 pandemic by using this increase in their services to help track and contain the spread of the virus (Fildes & Espinoza, 2020). Numerous companies worldwide have gone to various lengths to ensure people remain connected, whether it's offering unlimited minutes or different networking tools to help customers who are working from home. Simultaneously, these telecommunications companies, such as those in the United Kingdom and China, have also teamed up with the government and other organizations and provide data to aid in tracking the spread of the virus. Especially in China, its highly mobile population and its frequent data exchanged has proved an effective and essential tool in screening for infected individuals (Casey & Wigginton, 2020). Following this trend, many other countries are also likely to use cellular data in order to trace the number of viral infections within their communities.

The rise in telecommunication has been, expectedly, beneficial to telecommunication companies. With the exception of a few companies with sports-related media, the heightened use of their services has been more advantageous in comparison to struggling industries in the pandemic. This rise in telecommunication has also allowed for the emergence of new companies, such as Zoom, to establish a competitive stance in the telecommunication market as it becomes widespread across remote work and school programs (Novet, 2020). Many companies are also looking to make long-term investments in their networks, such as investments in 5G, in order to increase network reliability, resilience, and speed (Casey & Wigginton, 2020). At the moment, telecommunication companies must ask themselves important questions regarding how they might uphold the reliability of their networks, ensure a positive customer experience even when there is high pressure being placed on networks, and maintain consumer data privacy in the future amidst data collection for virus tracking.


**The Societal Effects**

While society has been responsible for the rise in telecommunication during the pandemic, this rise also has lasting implications regarding social changes in society. At home, many individuals have turned to different methods of telecommunication for working arrangements, alternatives for the classroom, as well as to simply stay connected with the world in isolation. Similar to the telecommunications companies, individuals of every age must now adapt as the world becomes more and more dependent on telecommunication technology.

*Remote Working Arrangements*
Unless deemed an essential worker, much of the workforce was sent home for safety reasons in order to prevent the spread of the pandemic. Those that were fortunate enough are able to work from home while others may have been laid off. While a small fraction of workers had already adopted the routine of remote work in different countries, a large portion of those sent home were forcibly experiencing this new arrangement for the first time. In reality, only about a third of occupations are able to work entirely from home, a rarity that has become a necessity in these dire times (Dingel & Neiman, 2020). Additionally, working from home requires certain infrastructure to which some might not have access. Resources such as electronic devices and secure network connection might seem mundane to some but can prove to be difficult to obtain for others. While certain workplaces have made sure to provide their employees with supplies and resources at home, many of those who have the luxury to work from home are often higher-income workers (Reeves & Rothwell, 2020).

As the increase in telecommunication continues, many predict that the pandemic will be a driving force of the trend towards long-term remote work for many. However, remote work and telecommuting have not become as widespread as many predicted when remote work technology first emerged (Vilhelmson & Thulin, 2016). Previously, many employers were uninterested in investing the resources, such as technology and management practices, that are necessary to operate a workforce with employees working from home. With the recent occurrence of the pandemic that forced many to work remotely and caused a surge in telecommunication needs, employers now realizethe benefits of remote work (Guyot & Sawhill, 2020). A recent survey revealed that one in five chief financial officers planned for at least 20% of its workplace to work remotely in order to reduce costs (Minaya, 2020). While the initial increase in remote work may have been seen as undesirable by companies, employers are now supporting their employees working from home which will cause telecommuting to adopt a heightened "new normal".

Although not to the same degree, the occurrence of a crisis has resulted in a spike in telecommuting in the past as well. In 2002 following the events of 9/11 and the anthrax attacks, interest in remote work increased in the United States after several government offices were forced to close (Tong & Schwemle, 2002). As well, workplaces in Christchurch, New Zealand also adopted the use of telecommunication and remote work after a series of earthquakes in 2010 and 2012 caused workplaces to close (Donnelly & Proctor-Thomson, 2015). During disasters such as these, schools and childcare were often limited or non-existent, leaving many individuals having to be both workers and parents. The COVID-19 pandemic is no different; with school closures, many individuals must tend to both their children and their work, making it much easier for those able to work remotely to perform both.

*Online Schooling*
As aforementioned, school closures are widespread across the globe in efforts to control the virus and preventing its spread. While online schooling had been present in certain school systems before, schools worldwide are now being forced to adapt to a new, remote learning system for their students. This change in learning style also requires students to adapt and presents itself with both disadvantages and advantages.

Technology is said to have negative effects on youth, and online schooling should not be exempt from that statement. In reality, it is not so much the schooling itself, rather the prolonged frequent use of technology required. Especially for younger children and adolescents, mental overload can result from an overuse of technology that may further disconnect individuals from others. As well, this disconnect may translate into less effective social skills when placed in non-virtual conversations. Prolonged use of technology for online school, similar to the effects of office jobs, may also force students to adopt a more sedentary lifestyle which can also be detrimental to physical well-being (Halupa, 2016).

Alternatively, the use of technology, specifically with regards to telecommunication, can have positive effects as well including academic achievement and increased feelings of social connection (Strasburger et al., 2010). On the contrary, this helps promote mental and physical well-being in which case the rise of telecommunication would exhibit a positive effect. Additionally, the advent of more online learning may provide students in rural areas with less access to in-person classes more opportunities to participate in diverse education experiences. Certain online classes also offer non-synchronous learning, during which students may learn at a pace that is comfortable and suited for themselves. The adoption of increased use of telecommunication for educational purposes can contribute to establishing self-regulation skills and goal-setting behaviours since both are necessary to succeed in a virtual learning environment.

*Social Connectivity*
In an already growing community of social media and technology users, the pandemic and the consequential isolation that followed demanded more methods to remain connected with friends, family, and the rest of the world. The world's active social media population has now reached 3.81 billion people in 2020, with many of those individuals being active on multiple media platforms (Clement, 2020). Looking at Facebook alone, the number of users on the social sharing platform has grown to 2.38 billion— an overwhelming increase since the last H1N1 pandemic in 2009 when Facebook only had 276 million users (Ortiz-Ospina, 2019). With telecommunication becoming more and more accessible to different communities around the globe, the increase in communication has also become the root cause of misinformation that has been

transmitted globally much like the virus itself. A survey indicated that adults between the ages of 18-29 in the United States are more likely to learn about the news through social media instead of print newspapers or news sites (Shearer, 2018). While the spread of misinformation will be explored further in a later chapter, this chapter focuses primarily on the increased connectivity gained through the rise of telecommunication.

While social media and the increased fluidity of communication has its drawbacks, the interconnectedness that it brings is essential for humans as social creatures. Social isolation and loneliness may contribute to rapid cognitive decline, increased negativity and depressive cognition, and is connected to a series of health concerns such as diabetes (Cacioppo & Hawkley, 2009; Social Isolation Among Seniors: An Emerging Issue, 2004). Especially in seniors and older adults, loneliness and isolation are affiliated with poorer cognitive function (Shankar et al., 2013). In these situations, especially since visiting loved ones in-person is not an option during the pandemic, telecommunication has enabled many to remain connected and preserve their psychological health.

**An Evaluation of Telecommunication in the COVID-19 Pandemic**

Telecommunication has undoubtedly become more prevalent in everyday lives, especially during times when face-to-face interactions are not possible. In light of isolation, telecommunication companies have benefited through the increase in use of their services and new partnerships with previous competitors to ensure the reliability of network connectivity for their customers. With regards to the general public, many individuals, including students, are now working remotely and benefitting from a more flexible environment that is likely to continue beyond the pandemic as well. Furthermore, telecommunication has allowed individuals to remain connected throughout isolation periods and help maintain their mental health. While this increase in telecommunication does provide implications with regards to data privacy and the long-term health effects of prolonged technology use, telecommunication, and its rise, is ultimately a necessary tool for society to persist throughout the pandemic.

# Chapter 8: The Internet and Misinformation

The COVID-19 pandemic has devastated the world, with millions being infected and hundreds of thousands dying in addition to the plethora of socio-economic consequences. However, many people may not realize that COVID-19 has, in fact, brought another pandemic with it; the pandemic of misinformation. Misinformation is defined as false, inaccurate, or otherwise misleading information, especially when it is meant to purposefully deceive an audience. The current situation has highlighted the influence that the internet and media have come to have on our population, with misinformation spreading fiercely online and having dangerous consequences. As a result of this, people are refusing vaccinations, taking remedies or drugs that are not proven to work, and ignoring public health directions, just to name a few of the ways in which the fight against COVID-19 is being harmed. When looking at how people consume information, it is unfortunately not surprising to see why this is the case. In the early months of the pandemic, posts from the World Health Organization (WHO) and the Center for Disease Control (CDC) only received a few hundred thousand engagements whereas hoax and conspiracy theory sites have amassed over 52 million engagements (Mian & Khan, 2020).

The power of misinformation is unlike anything else and has proven time and time again to be fatal for many; COVID-19 is no exception. For example, after groups in Denmark and Ireland broadcasted testimonies of the human papillomavirus (HPV) vaccination harming young girls, national immunization rates for the virus in Denmark fell from over 90% in 2000 to under 20% in 2005 (Larson, 2018). Additionally, this is not the first time that misinformation has damaged efforts to fight an epidemic. In the early years of the HIV/AIDS epidemic, rumours and conspiracy theories about the virus spread to such an extent that it influenced the South African Government to deny the legitimacy of HIV in 2004 and consequently deny access to

Antiretroviral Therapies to thousands of mothers (Mian & Khan, 2020). It is estimated that over 343,000 people died as a result (Bateman, 2007).

**The Power of the Internet**

*Social Media*

Social media may be the most powerful tool available for the communication of information, but this means that it can have both positive and negative consequences. Although it provides the unique and invaluable opportunity to connect with one another across the globe, it also gives everyone, including those without proper expertise on certain subjects, the ability to make claims which can then be spread without any sort of verification. This means that it is also one of the most effective ways through which false and fear-mongering information can spread to millions of people. The primary issue is that misinformation spreads faster than information from credible sources, thus "damaging the authenticity balance of the news ecosystem" (Tasnim et al., 2020). People tend to act and operate based on the information they receive, and when this information is false, those actions can be detrimental. In fact, many instances of racism and xenophobia, mass panic, and misleading rumours regarding COVID-19 can be largely attributed to social media (Depoux et al., 2020).

Unfortunately, the prevalence of social media and its role in the pandemic differs greatly from previous pandemics such as MERS and H1N1. This is because, unlike before, rather than members of the public simply consuming inaccurate and misleading information, they are now actively spreading and even creating it, according to Lee Kum Kee, a Professor of Health Communication at the Harvard T.H. Chan School of Public Health. He goes on to state that "the sheer volume of COVID-19 misinformation and disinformation online is 'crowding out' the accurate public health guidance, 'making [their] work a bit more difficult'" (Pazzanese, 2020). Claims such as this are backed by scientific studies, as demonstrated by one that looked at 673 different tweets relating to 14 different trending hashtags and keywords about COVID-19. Of these tweets, 153 included misinformation and 107 included unverifiable information, while a staggering 448 tweets were made by informal individuals or groups (Kouzy et al., 2020). With so many of the most popular tweets containing false or unverifiable information, it is no surprise that misinformation can spread so rapidly. Moreover, an average user may see the same false or misleading information coming from different accounts and groups. Its frequency may then give it more credibility from the perspective of the user and increase the likelihood of them sharing the post.

*The Web*

In this day and age, the internet provides anyone with the opportunity to publish anything that they wish online for others to see, meaning that it is impossible to inherently trust websites. For example, a study was conducted that analyzed the first 110 google search results for "Wuhan Coronavirus." This was done using the Health on the Net Foundation Code of Conduct (HONcode), the Journal of the American Medical Association (JAMA) benchmark, and the DISCERN Instrument (Cuan-Baltazar et al., 2020). HONcode aims to help standardize the reliability of medical and health information available on the World-Wide Web and provides certification (The Health on the Net Foundation, 2019), and JAMA uses 4 core standards to evaluate websites: authorship, attribution, disclosure, and currency (Cassidy & Baker, 2016). DISCERN is "an instrument, or tool, that has been designed to help users of consumer health information judge the quality of written information about treatment choices" (DISCERN, n.d). Of the 110 sites analyzed, only 1.8% had the HONcode seal, 39.1% met none of the standards set out by the JAMA benchmark and only 10.0% met all 4 standards. Additionally, the DISCERN instrument evaluated 70.0% of websites as having a low score and did not give a high score to any website (Cuan-Baltazar et al., 2020). This evaluation demonstrates the ease with which misinformation can be spread across the internet, as the first 110 sites have shown little to no reliability in the information that is being conveyed.


*Media Outlets*

The media also plays a large role in the spread of misleading information and information that only focuses on one aspect of the larger picture, as people generally rely on news outlets which gives them a large degree of perceived credibility. However, these media outlets all have their own biases and agendas, which may motivate them to publish information that favours a particular side, and are often looking to increase views and ratings. Unfortunately, humans are generally more attracted to negative news, meaning that it is no surprise that these types of stories are the most prevalent in the media (Trussler & Soroka, 2014). In the context of COVID-19, this can mean more of a focus on negative events and disasters relating to the virus which will only serve to spread panic. Additionally, individuals who are normally more fearful or anxious generally pay more attention to threat-related information, and the volume of COVID-19 coverage coupled with media bias results in plenty of opportunities for those with a higher fear of contamination to focus on the threatening aspects of the outbreak (Clark, 2020). Regardless of the individual, however, the focus on negative stories and the subsequent mass panic has led to a slew of consequences. For example, many people rushed to stock up on resources such as food and toilet paper, leaving shelves empty. This was unnecessary and harmful, as it led not only to vulnerable individuals such as the elderly not being able to purchase necessities but also to crowding in stores which may have contributed to the spread of COVID-19. Similar consequences were seen when CNN published a report on March 8, 2020, stating that a northern region of Italy will be locked down to contain the spread of the pandemic (John & Wedeman, 2020). This was hours before any official communications from the government and led to citizens panicking and overcrowding trains and airports, potentially aiding the spread of COVID-19 (Cinelli et al., 2020). Overall, it is clear that media outlets have a profound impact on populations, but ulterior motives such as bias and the pursuit of ratings, the level of power that they hold can, unfortunately, harm populations on a large scale and increase the devastation caused by the pandemic.

## The COVID-19 Vaccine

The anti-vaccine movement is by no means a new concept, but with a COVID-19 vaccine being currently in development, it is no surprise that this is one of the most significant opportunities for misinformation and public pushback. In addition to the general anti-vaccine arguments that are often made by opposing groups, such as the claim that they cause autism and a slew of other health complications, the COVID-19 vaccine has come with some rumours of its own. Some of the most prominent pieces of misinformation currently making their way across the web are the rumours surrounding Bill Gates' involvement with the virus, with individuals and groups claiming that he had created the virus and patented the vaccine or that he would use these vaccines to track and control individuals (Ball & Maxmen, 2020). Another demonstration of just how widely misinformation can spread comes from biohackinfo.com, which on March 19, 2020, falsely claimed that Bill Gates was planning to use the vaccine to inject a tracking chip into patients. Within 2 days, attention was turned towards a YouTube video about the matter which has since been viewed over 2 million times and the idea eventually reached Roger Stone, a former advisor to President Donald Trump, who discussed the matter on a radio show and claimed he would never take a vaccine that was funded by Bill Gates. That interview was covered by the *New York Post* in an article which was then liked, shared and commented on by nearly 1 million people. Soon after, many high-level profiles with large followings also commented on the matter, further spreading the misinformation (Ball & Maxmen, 2020). This example perfectly demonstrates how although the roles of social media, the internet, and media outlets individually were discussed in this chapter, they all work together to perpetuate the spread of misinformation. Something as small as a claim on the internet from a source with no credibility was suddenly reaching millions via social media and led to several prominent figures and media outlets sharing the issue without any idea that it was false. Unfortunately, misinformation seems to have a significant impact on the general population as a survey conducted by The Associated Press-NORC Center for Public Affairs Research found that only about 50% of Americans would agree to get the vaccine when it is released. Additionally, while another 31% claimed to be unsure, that still leaves 1 in 5 Americans that would refuse the vaccine altogether (Neergaard & Fingerhut, 2020). This ultimately may prove to be the worst impact of misinformation, as without adequate vaccination, the virus may continue to spread.

## Combatting Misinformation

In a time where misinformation is abundant and wreaking havoc on efforts to fight the coronavirus, the duty falls on everyone to limit its spread. There are measures that can be taken at every level, ranging from the federal government to the average individual. On a broader level, one of the first measures that must be taken is to provide healthcare workers with the current and accurate information regarding COVID-19 which will not only allow them to provide better care, but also to communicate this information to patients (Tasnim et al., 2020). In order to achieve this, a system through which this information can be easily accessible to healthcare workers will most likely need to be developed, and it can even be made accessible to the general public. Additionally, citizens may place more trust in the information they hear directly from their doctors which will hopefully reduce the spread of misinformation. Another step that can be taken is for major media and information companies, including social media, to coordinate with one another and develop a method to tag posts containing misinformation and warn users to be cautious with how they interpret the post (Tasnim et al., 2020).

The final measure which can potentially be taken on a broader level is for governments and media companies to remove all misinformation from the internet altogether and ensure that only scientifically sound information is made available to the general public (Tasnim et al., 2020). Although this may be the strongest tool for removing misinformation, it also has the most potential problems. The most significant concern with this is that it can be considered censorship and clashes directly with freedom of speech. Freedom of speech is one of the fundamental principles of democracy and the argument can be made that this should not be undermined and that even if false, people have the right to share what they believe. Additionally, it would require citizens to trust and give more control to the government, an idea to which many citizens may be opposed. This also may present an opportunity for those who are against the measures being taken by the government to stop the spread of COVID-19 to further convince others of their viewpoint, as they may claim that the government is doing this as a "cover-up". Unfortunately, the unorthodox nature of censoring information in a democratic nation, as this is not done for misinformation relating to any other topic, may give credibility to this claim for some citizens which can result in greater public backlash. If the decision is made to censor misinformation, there will surely be a great deal of controversy and outcry regardless of the benefit it brings.

Although the aforementioned measures must be taken on a societal level, there are things that every individual can do to ensure they minimize the role they play in the spread of misinformation. First and foremost, it is important to ensure that the only sources that are being consulted are credible ones. For COVID-19, this can include government and public health websites, university pages such as that of John Hopkins, or research articles found through a variety of online databases. Additionally, individuals must cross-check any information, regardless of whether or not the source is credible, with other credible sources to ensure consistency in what is being reported. Any information that is presented by a news outlet or individual should be verified first before it is reposted (The Johns Hopkins University, 2020). The hope with these measures is that if everyone were to take a few minutes to research a topic or claim for themselves before reposting information, the spread of misinformation could be significantly hindered.

Overall, it has become evident that there are currently 2 pandemics descending on the world; COVID-19 and misinformation. This has led to a plethora of negative consequences that have not only undermined the fight against the virus, but in some situations has also actively harmed or killed individuals and populations. Misinformation on the internet is by no means a new phenomenon, but with the number of internet and social media users constantly rising, it is poised to be more dangerous than it has ever been before. In order to reduce its spread and to subsequently reduce the spread of COVID-19, every individual must play their part and actively seek out credible, scientifically sound information while ensuring that false or misleading information is not shared and reposted.

# Chapter 9: Physical Health Effects of Technology

**Introduction**

As many parts of the world have instituted lockdown procedures to help curb the spread of COVID-19, many of the usual entertainment streams people flock to have shut down, like parks, playgrounds, bars and restaurants (Davidson, 2020). This has led many to seek other forms of entertainment, usually resorting to their technological devices to pass the time (Koeze & Popper, 2020). Furthermore, many services that help individuals stay active have been moved online; a drastic shift from the norm (Roberts, 2020). These changes have had serious impacts on people's physical health, which will be discussed in this chapter.

**Impact on Children**

While the effects of technology on physical health discussed in this chapter affect all members of the population, they are especially problematic for children. This is because children's brains are still in the process of development and thus may be more sensitive to technological overuse in comparison to adult brains (Johnson, 2020). Some of these effects may manifest themselves in different ways and hinder a child's development and learning. These manifestations include low academic performance, low creativity, lack of attention, obesity, delays in social development and delays in language development (Johnson, 2020). Furthermore, a study of teens between the ages of 15 and 16 discovered that those with high digital media use had an increased probability of developing ADHD symptoms (Ra et al., 2018). These effects highlight the increased impact technology has on the physical health of children and thus, a focus will be placed on this demographic in this chapter.

**Screen Time-related problems**

The pandemic has forced many to stay indoors and spend more time on screens. In fact, according to a study done by Axios, most children aged

6 - 12 are reporting spending at least 50% more time in front of screens daily (Fischer, 2020). This has been supported by the increase in internet traffic across the board, with kids apps experiencing nearly 70% more traffic (Bleyleben, 2020), gaming apps seeing a nearly 20% increase (App Annie, 2020) and tablet/phone traffic almost doubling (Fischer, 2020). This increase in screen time can cause a wide variety of physical health-related issues which could have devastating effects over time.

*Eye-strain and vision-related problems*
One significant impact of spending more time on screens is an increase in eye strain and vision problems, which is referred to as "computer vision syndrome" (Forster, 2020). Our eyes were not designed to spend many hours staring at computers reading tiny text, and an increased reliance on doing so can lead to a wide variety of vision-related problems. These effects include increased eye strain, fatigue and headaches. Furthermore, if continued for a long period of time, this can also leave eyes red, parched and with a gritty feeling (Forster, 2020). According to Scott Drexler, an Assistant Professor of Ophthalmology at the University of Pittsburgh, Computer Vision Syndrome-related effects have become "incredibly common" during COVID-19 with blurry vision and headaches being the "most common complaints" during consultations (Forster, 2020).

In children, these effects are especially drastic, since an increase in screen time and a decrease in time spent outdoors may actually place them at a higher risk of developing myopia (Shih & Killeen, 2020). Myopia occurs when the eye's ability to focus is too strong, which causes light to focus in front of the retina instead of on it, leading to the creation of a blurry image (Turbert, 2019). According to a review of 25 years of research, working up close to a screen, like reading or using a tablet, increases the chances that a child develops myopia (Huang, Chang, & Wu, 2015). Furthermore, researchers in Ireland found that greater than three hours of screen time each day increased the odds of myopia in schoolchildren (Harrington, Stack, & O'dwyer, 2019). While this can usually be fixed with glasses or contacts, according to a study done by Dr. Yasushi Ikuno, myopia can place children at greater risk for numerous problems down the road like macular degeneration, retinal detachment and glaucoma (Ikuno, 2017). These findings indicate the importance of reducing screen time during COVID-19 to prevent the development of potentially long-term consequences. While the move to remote work might necessitate the frequent use of technology, some strategies can be employed to reduce damage to vision (Forster, 2020). This includes the "20-20-20 rule" which urges frequent screen-users to take a break every 20 mins to look at an object 20 feet away for 20 seconds (Forster, 2020). Furthermore, individuals can also purchase glasses with blue light filters, which help reduce the eye strain and other associated effects (Forster, 2020). By employing these techniques, the impacts of screens on vision can be mitigated.

*Posture-related problems*
Another side effect of increased screen time is an increase in posture-related health concerns. This includes pain and straining in the neck, back, and shoulders, which occur because people do not follow proper ergonomics procedure while using their technological devices. Other problems may also be experienced which stem from this bad posture. This includes headaches, as muscles in the neck are strained, placing pressure on the head and fatigue, as the body must work harder to maintain an upright posture (La Trobe University, 2015). In fact, according to a study done at Boston University, 50% of university students stated that they have experienced or continue to experience back and neck pains (Murphy, 2011).

One of the most prominent areas where this pain is felt is the neck. The human head, on average, weighs more than 20 - 30 pounds, which is well-supported in an erect position by the many muscles in the shoulder and neck (McCabe, 2020). However, when an individual constantly leans forward while staring at a computer screen, these muscles become stressed and inflamed, which leads to neck pain (McCabe, 2020). Another area where this pain is most often felt is the back(Vad, 2020). If the back is not in its natural position and is unsupported, this can cause the "loads on [an individual's] spine to disperse incorrectly." This weakens the tissues in the lower back and causes back muscles and joints to be pushed beyond their limit, resulting in pain (Vad, 2020). While avoiding screen time altogether during quarantine is difficult because many schools and workplaces function through the internet, measures can be taken to maintain proper posture and avoid these effects. This includes keeping the head upright, back straight and shoulders back to enable proper back and neck support (Mayo Clinic Staff, 2019). These changes must be implemented both by actively adjusting posture to meet these conditions and also ergonomically designing workspaces to support these changes.

*Wrist Pain and Injuries*
Wrist pain and injuries are another common consequence of the over-use of technology. These effects have become increasingly exacerbated as a result of the shift towards online work and school platforms stemming from the lockdown procedures that have been initiated (Koeze & Popper, 2020). This pain and the resulting injuries occur because of bad wrist posture when using computers, phones and other technological devices (Premier Orthopedics, 2015). The muscles, tendons and nerves in the wrist are repeatedly placed in a compromising position, which leads to damage that builds up over-time (Hecht, 2017). This can lead to a wide variety of symptoms, such as pain, stiffness, throbbing, tingling and cramps (National Health Service, 2018).

Furthermore, over the long term, more serious conditions like Carpal Tunnel Syndrome and wrist arthritis can also occur. Carpal Tunnel Syndrome is a condition where, because of repetitive stress caused by bad posture, bones and ligaments place pressure on the median nerve, which leads to symptoms like numbness, tingling and weakness in the hand/arm (Shiri & Falah-Hassani, 2015). This occurs in the carpal tunnel, which is a narrow passageway on the palm side of the hand through which the median nerve passes. Wrist arthritis is a condition where the joints in the wrist become inflamed. This occurs because repeated bad wrist posture causes damage to the cartilage between the bones in the wrist. The cartilage is gradually degraded, which causes the bones to rub against each other and leads to pain and stiffness at the joint (Belliveau, 2017).

To avoid symptoms like this, both during the pandemic and after, individuals should seek to maintain proper wrist position during their use of technological devices. While typing, this involves placing the wrist in a neutral position with the thumb in line with the forearm and the wrist slightly back. It also recommended to alternate between resting

60

on the pads of the hand and raising the wrist up to prevent repetitive stress in one position (Healthwise Staff, 2019). Furthermore, it was found that typing on smartphones with both thumbs is stressful on the wrists. Thus, it is recommended that individuals place their phones on tables while texting as much as possible to reduce damage to the wrist (Ventura Orthopedics Staff, 2018).

*Sleep*

One of the biggest side effects of increased technology use during quarantine is its impacts on people's quality of sleep. According to Donn Posner, a member of the T.H. Chan School of Public Health, the current, COVID-19 situation is the "perfect storm for sleep problems" because of the disruptions to people's daily routines, increase in night-time technology use and increased stress. Of the three, he notes that the effects of an increase in night-time technology use is especially devastating (Simon, 2020). This is the case for a few reasons;

First, the blue light emitted by screens from tablets, computers, smartphones and televisions can restrict the production of melatonin, the hormone that regulates an individual's sleep/wake cycle and circadian rhythm (National Sleep Foundation, n.d.). Furthermore, when technology is used to check emails, watch movies or surf the web before bed, the mind is kept engaged and alert. This makes it difficult for the individual to quiet their minds while falling asleep, which can impact the length and quality of sleep (National Sleep Foundation, n.d.). Finally, even if technology isn't being used by an individual, keeping it within reach at night can disturb an individual's sleep. This is because of notifications from texts, emails etc. that come in at night, which can cause the individual to wake up at numerous points throughout the night (National Sleep Foundation, n.d.).

To mitigate the impacts of technology on sleep, a few different strategies are advised . First, individuals should stop using technology at least 30 minutes before going to sleep. Furthermore, electronic devices should be placed in a different room or with notifications turned off so that sleep goes on uninterrupted. Finally, individuals can download applications on devices that reduce the blue light screens emit to achieve better sleep (Robinson, 2019).

*Reduced Physical Activity*
Technology has also caused a decrease in physical activity during the pandemic, particularly in children. Before the pandemic, many schools across the world had measures to keep kids active, including lunch breaks where students were encouraged to go outside, gym lessons and the incorporation of activity into classroom lessons (Kohl & Cook, 2013) . However, since the shutdown of schools in many places around the world, many kids have been stuck at home (Koeze & Popper, 2020). While they have the option to go outside and play, video games, Youtube, Netflix and other technological platforms are seen by many children as a better, more entertaining alternative (CYCADSG, 2019). In fact, the video game industry has thrived during the pandemic, with many companies reporting 25%+ increases in usage year-over-year (Smith, 2020).

*Impact of virtual home workouts*
In addition to an increase in screen time, quarantine procedures worldwide have also caused a shift to at-home workouts (Benveniste, 2020). In fact, 18% of Americans indicated in a study done by Pew Research that they had participated in an online fitness class or followed a pre-recorded online workout at home (Vogels, 2020).This has caused many to rely on technological tools to craft an effective workout regimen. This ranges from relatively low-tech options such as virtual workouts hosted by fitness trainers both live and pre-recorded, to MIRROR, an interactive, at-home fitness system that looks like a full-length mirror. These products have reportedly seen a boom in sales, almost "doubling pre COVID-19 amounts"(Benveniste, 2020). Many of these training guides are often offered free of charge on social media platforms, where they can reach a wide variety of individuals (Suciu, 2020).

By providing guidance on a way for many around the globe to practice regular exercise even while training facilities and gyms remain closed, these virtual platforms have proven to be crucial in helping individuals remain physical during the pandemic. Regular physical activity has been shown to boost the immune system, which helps fight off infection (Centre for Disease Control, 2020) . Furthermore,regular exercise helps reduce the risk of heart disease, strengthen bones and muscles and reduce risk of falls. It also helps stave off chronic conditions, allowing individuals to live longer, fuller lives (Centre for Disease Control, 2020). In this way, technology has had a positive impact on individual's health and wellbeing.

# Chapter 10: Psychological Effects of Technology

By now, it may be evident that technology is one of the defining factors of this pandemic that makes it vastly different from those in the past. Technology has helped in developing accurate diagnostic technologies, vaccines, and preventative measures such as masks and ventilators. In addition, technology has facilitated the spread of information and advocacy. The forwards and posts on platforms such as Instagram and WhatsApp have certainly made larger populations of people aware of the severity of the pandemic as well as informed individuals on how to stay physically and mentally healthy. News platforms have their content readily available online, which also aids in informing the entire population on the social distancing conventions. While the advocacy taking place during the pandemic is beneficial in many ways, many journalists have been using the term 'infodemic' to describe the spread of misinformation through technology that has occurred alongside COVID-19 (Harris et al., 2020). The spread of such information has the potential to generate fear and panic, which has psychological impacts. Therefore, while the use of technology during this pandemic has allowed people to stay safe, it also has its downsides. From people spending excessive amounts of time on social media, to individuals panicking about the current state of affairs and engaging in impulse buying, it is evident that the use of technology must be examined from a more critical lens. There is a dichotomy in the ways in which technology may impact our mental health. Thus, this chapter will focus on the various ways in which technology has psychological impacts.

## Fake News and Mental Health

The outbreak of the novel coronavirus was unsettling for the public and was associated with unclear advice from the government and misguided conspiracy theories that seemed to pose additional health risks. This unfortunate spread of so-called 'fake news' has led to a new pandemic of misinformation (Harris et al., 2020). In the past decade, fake news has become more prominent due to online social media. Studies have shown that believing fake news related to COVID-19 perpetrates feelings of anxiety, stress and depression (Wright, 2020). Some examples of fake news include the belief that a Harvard professor sold the coronavirus to China. The belief of these conspiracy theories is more common than may be perceived. In fact, almost half of Canadians believe at least one popular conspiracy theory, namely one that states that the coronavirus was made to be a bioweapon made in a Chinese lab and released to the general population. A popular vehicle for these conspiracy theories has become memes shared through Twitter, Snapchat, Instagram, TikTok, and Whatsapp. The consumption of fake news and conspiracy theories through these various platforms is one of the reasons why 66% of Americans are constantly stressed about the future (Erdelyi, 2020). The ways in which fake news stories come to be is much different in the current global climate in comparison to prior pandemics. Misinformation is no longer passively spread. Rather, it is constantly being disseminated and created, resulting in decreased feelings of trust towards public health agencies. Challenging times like these make feelings of certitude so rare, and the spread of fake news makes it even more difficult to find credible and reliable information. The lack of consensus-oriented information adds to the confusion among people. Furthermore, mainstream media coverage generates more fear and anxiety as the individuals relaying the news have little to no training on public health matters and struggle to accurately convey scientific information. If these news sources that were once the golden standard are no longer reliable, individuals begin to lack a sense of trust in these sources generating even more confusion (Pazzanese, 2020). Facebook, Twitter and Youtube have all increased their efforts in taking down COVID-19 misinformation, but their efforts have fallen short. This is because battling misinformation is a harder task when there is not that much information about the virus in the first place. To combat this issue, governments have been working with social media companies and disinformation specialties to first identify the extent of misinformation, and to begin launching public information campaigns to help people fact-check their sources (Pazzanese, 2020).

## The Spread of Misinformation

Another branch of the 'infodemic' during COVID-19 is the spread of online hate, specifically towards people of Asian descent. Civil rights groups have identified an upsurge of racist remarks on social media platforms as a result of comments made by powerful figures such as Donald Trump. Terms such as 'kungflu' have perpetrated the online hatred and distrust towards China. It has been shown that due to the racist comments made online, people who look Chinese are at increased risk at experiencing racism similar to the discrimination faced by Muslims wearing turbans. This type of racism is important to discuss as it has a profound psychological impact. There is a definitive link between mental illness and experiences of discrimination as it is a major stressor. It leads to feelings of not being good enough, and may even result in symptoms of post-traumatic stress disorder (PTSD). Data from the University of Southern California's Center for Economic and Social Research shows that 18% of Asian Americans have reported encountering instances of discrimination during the pandemic. Prior to the pandemic, it was shown that white Canadians experienced a greater decline in mental health and more mental illness symptoms than east-Asian Canadians. However, during the pandemic, it has been shown that East Asian Canadians have poorer mental health. It has been suggested that instances of acute discrimination can explain over 20% of the East Asian-white mental health gap (Wu et al., 2020). Thus, since technology, specifically news and social media platforms, have been a vehicle for racism during the pandemic, it may be concluded that the use of such platforms has a negative psychological impact, more so for individuals of Asian descent.

## Psychological Consequences of Social Media

Not only has technology allowed for the spread of misinformation, it generally has a vast array of psychological consequences. Studies conducted early on the pandemic in China have suggested that social media exposure is correlated with an increased prevalence in mental health problems (Gao et al., 2020). When evaluating the effects of social media on mental health, many factors must be considered. First, past research has indicated that mere exposure or time spent on social media is an indicator of mental health. Additionally, the significance of social media to an individual determines the ways in which they consume information and can indicate how much it affects their mental health. Studies have shown that *how* social media is used is a more important predictor of mental health than merely exposure. However, with the recent uprising of TikTok as a popular social media platform, many of these issues relating to mental health have been tackled. TikTok is a Chinese video-sharing social networking service. It has over 500 million monthly users. Since being forced into isolation, people around the world have been using the app to connect with others. Educators have even tried to harness its popularity and use it as a way for children to learn. This app is such a sensation because it allows people to utilize their creativity to create compelling content. The length of these videos can range from 3 to 60 seconds, and have led to many conversations being started about personal topics such as depression, suicide, eating disorders, racism, and social distancing. Before TikTok became popularized, one of the main concerns with regards to social media and mental health was the fact that people project a 'perfect' image of themselves, which is often not really who they are. This can lead individuals who view such content to constantly compare themselves to an unrealistic version of perfection. However, on TikTok, many popular users have made it their mission to step away from this need to be perfect, and rather, share their stories. For example, Rianna Kish, a high school student from Alberta, has gained a massive following base through posting videos about self-love and body image. She is in the process of recovering from an eating disorder, and her positivity and truthfulness online has helped numerous teens in a similar position feel supported. Humor has also been a crucial tool in helping people, especially adolescents through this difficult time, and TikTok has been an outlet for people to share their views on the pandemic in a lighthearted manner. Further, the vast number of coronavirus memes have helped people of all ages cope with the dread associated with the pandemic. Recent memes have poked fun at the fact that people were hoarding toilet paper and feelings of going stir crazy in quarantine (Gao et al., 2020).

## Retail Therapy during COVID-19

Another way in which technology has had a profound psychological impact is through the change in scenery with respect to the retail industry. Canadian consumers have shifted to online shopping during the lockdown, at which point sales surged 99.3%. E-commerce sales have also had a record $3.9 billion (Italie, 2020). This is a result of panic buying, over-buying and even emotional buying. There have been cases where despite being in a recession, families have purchased expensive furniture, toilet paper, coffee, and other random household items. This has caused many of the traditional buying patterns to be tossed in the air, causing much confusion among economists. Many have claimed that these online purchases have made them 'instantly' feel better. It has been shown to reduce negative feelings and increase overall happiness, although this relief is short-lived (Italie, 2020). Online shopping has been an escape for many people as it is a mindless, relaxing activity (Yarrow, 2013). At the same time, it is also an addictive behaviour and is a short term solution to these feelings of anxiety during the pandemic.

## Access to e-Mental Health Services

Technology has also changed the way mental health services can be accessed, which is especially important during COVID-19. For example, there are several mobile mental health services that operate through text messages – a simple but effective concept. There are more sophisticated apps that also collect information on a user's typical behaviour and provide a signal when there is a change in this behaviour. There are services available to connect individuals to peer counsellors or health care professionals. Specifically, telepsychotherapy is the best known form of E-mental health. The power of digital technologies must be harnessed to improve access to and effectiveness of treatment. During COVID-19, the feelings of anxiety and loneliness have increased the need for mental health services. Thus, teleconsultations have become the new normal (Vial, 2020). However, prior to the pandemic, there was already a lack of tele-mental health solutions that could be used for psychiatrists, but a surge in non-scientifically based mental health apps. The success of these apps is not necessarily tied to the pandemic, but results from the growing demand for this type of healthcare. Headspace and Insight Timer have been some of the most commonly used wellness apps during this time. They provide free guided meditations that have introduced people of all ages to some strategies that promote mental well-being (NIMH, 2020). In addition, there are several online anxiety and depression support groups online, such as Turn2Me, which hosts free sessions run by qualified professionals. This network has also offered sessions to help people specifically cope with the dread of COVD-19 (Kindelan, 2020).

It's evident that COVID-19 came as a shock to most of the population. An event that causes such a drastic change in our way of life is bound to have psychological impacts. The lockdowns that have been initialized worldwide undoubtedly resulted in people having intense feelings of loneliness, anxiety, and a general altered worldview. This led many people to turn to technology as an escape and a coping mechanism. The use of technology in the form of online shopping and several social media platforms has the ability to cause heightened feelings of depression and loneliness. On the other hand, platforms such as TikTok have been utilized to spread positive messages, and the increase in mental health resources online has allowed individuals to adequately cope with the current situation. In times of distress and confusion, this pandemic has shown us that we must have faith in our global community to find solutions to the problems we face, from the physical spread of the virus to the psychological problems associated with social distancing.

# Chapter 11: Technological Demands on the Healthcare System

As technology continues to improve and open endless possibilities for development, healthcare has similarly followed in light of its recent advancements. Technology is an essential component of modern medicine and has helped improve and save the lives of many. Especially during a global pandemic, the need for technology, not only in terms of sophistication but also quantity, has never been more apparent. Global shortages of crucial devices and apparatus have since galvanized a variety of institutions into action to research and develop refined alternatives while simultaneously attempting to bridge the shortage gap with existing technology. However, the repercussions of healthcare systems inadequately equipped to battle the novel virus are widely evident in countries around the world. This chapter will explore the different technologies involved in the fight against COVID-19, its demands from patients, and finally, the disparities in access to these coveted technologies between different kinds of communities.

## An Introduction to Technology in the Healthcare System

A variety of different kinds of technology are present in the healthcare system. These devices, machines, and tools are important to both the patient and the healthcare provider and are used to either help diagnose, treat, support or protect either the patient or the provider.

### Technology for Patients

Technologies used on patients primarily consist of diagnostic and treatment technology. Diagnostic devices are used to determine the root cause or nature of a medical phenomenon, oftentimes regarding an illness. Examples of these devices include magnetic resonance imaging, temperature sensors, or x-ray scans. In the case of COVID-19, testing kits are utilized to identify individuals who have been infected by the virus. However, certain unreliable kits may not produce accurate results and therefore lead to false positives or false negatives (Ramdas et al., 2020).

Alternatively, treatment technology is used in attempts to address or solve the phenomenon previously identified by the diagnosis. Treatment options can include prescriptions for different kinds of medicine or surgeries. Unfortunately, with regards to COVID-19, there are currently no known effective treatments. While this is true, devices that offer supportive therapy, such as ventilators, help patients breathe and maintain essential bodily functions in hopes of recovering (Manthous & Tobin, 2017). It is important to not confuse ventilators as a treatment device, since ventilators do not actually cure the patient of the infection.

*Technology for Healthcare Providers*
Technology that is primarily used for healthcare providers, especially in light of the COVID-19 pandemic, is protective technology. While equipment such as gloves, medical masks, face shields and gowns do not have any electrical circuitry, they are still devices vital to protecting healthcare workers on the front lines. These devices are life-saving to healthcare providers treating patients infected with the COVID-19 virus and help prevent its spread (WHO, 2020c).

Other technological advances in recent years, such as augmented and virtual reality, artificial intelligence and telemedical robots are also helping frontline workers tackle the novel virus. Many of these devices are able to aid healthcare workers with their mental health and psychological well-being. Healthcare workers can also receive work training through these devices, ensuring that an experienced healthcare provider can help a patient instead of spending time training new employees. The onset of telemedicine also allows non-COVID-19 related patients to receive counselling and consultations without either the patient or the healthcare professional being at risk for contracting the virus (The Top 5 Practical Digital Health Technologies in the Fight Against COVID-19, 2020).

**Technological Demand (COVID-19 Related)**

The onset of the COVID-19 pandemic has undoubtedly created an increase in demand for technology within the healthcare system. As hospitals and institutions rush for life-saving resources, many regions are still experiencing a shortage of essential equipment and technology. The demands placed upon the healthcare system have outpaced the system's ability to provide these coveted medical supplies.

*Diagnostic Technology*
Devices used to diagnose illnesses are equally as important as the devices used to treat those same illnesses. However, limitations to the diagnoses that can be performed can have adverse effects on a population in times like a pandemic. Tests for COVID-19 are being increasingly demanded with insufficient resources to support this demand. One reason for the shortage is the diminishing supplies of essential components needed to perform these tests. Chemical reagents, which are required by laboratories to complete COVID-19 tests, are now in short supply and struggling to keep pace with the growing demand for tests (Malone, 2020). As well, a lack of trained personnel to complete these tests also contributes to the backlog of COVID-19 diagnoses. In Ontario, over 500,000 medical tests were performed daily, even before the pandemic (Weikle, 2020). Now with the surge in COVID-19 testing, laboratories are becoming overwhelmed. Countries worldwide are experiencing shortages of tests for the novel virus and are therefore unable to identify infected individuals. This leads to an inability to track the virus, and later, potentially the inability to properly contain its spread.

Due to its importance, this surge in demand for testing kits, and subsequently its shortage, has also increased the demand for technological innovation with regards to diagnostic devices for COVID-19. This pandemic has mercilessly demonstrated the shortcomings of existing diagnostic devices and medical testing infrastructure, as seen through the growing shortage of tests that are available to patients. To combat this, various biotechnology companies, institutions, and organizations are working to develop unprecedented methods to help test and diagnose individuals using advanced technology such as the recently discovered CRISPR gene-splicing tool (Sheridan, 2020). Likewise, governments are supporting and nurturing projects such as these through generous funding in efforts of rapidly producing a reliable and dependable alternative to current testing. In the United States, Congress has more than doubled the budget for the National Institute of Biomedical Imaging and Bioengineering with an additional USD $500 million (Sheridan, 2020).

*Treatment Technology*
While there are currently no known effective treatments for the COVID-19 virus, different kinds of medical technology can still be used to help support a patient in hopes of recovery. One of these devices is a ventilator and is considered to be supportive therapy for patients in staying in the intensive care unit (ICU) after being hospitalized for COVID-19. While not all patients infected by COVID-19 will require hospitalization, approximately one in every five infected individuals will develop severe difficulties with breathing and require to be admitted to the ICU (WHO, 2020d). However, when the pandemic first began infecting countries and communities around the globe, a substantial number of ICU beds were occupied by patients with critical conditions other than the COVID-19 virus. By now, operation rooms, waiting rooms, and even parking lots have been refashioned into alternative ICU spaces due to the overflow of COVID-19 patients (Grady, 2020). While the space for more beds has been exhausted, the number of ventilators available for patients is far fewer than what is needed. Some hospitals were forced to decide which patients would be fortunate enough to be admitted to the use of a ventilator while others were left to their own devices due to the limited resources available. Countries in which governments had previously invested in additional ICU resources were subsequently less affected than those that did not (Saunders, 2020).

Similar to diagnostic technology, this shortage of life-saving devices has galvanized institutions and created an increased demand for technological advancements and innovation. Many patients who are waiting for test results may also need to use ventilators and therefore limit the number of available ventilators for patients who have already been diagnosed (Ranney et al., 2020). Devices such as modified versions of traditional ventilators are being developed to increase the ventilators available for COVID-19 patients (Kunzmann, 2020). As well, a number of health institutions are also racing to develop a reliable vaccine as a cure to the current pandemic. The rapid development of vaccines against the virus is aided in part by previous knowledge regarding the roles of different proteins in the virus and has allowed its development to reach "Phase 1" in a period of 6 months; this process would typically adopt a timeline of 3 to 9 years (Heaton, 2020).

*Personal Protective Equipment*
While the health and well-being of patients are important, the safety of healthcare workers is equally as significant in the fight against the COVID-19 virus. With a shortage of personal protective equipment (PPE), medical personnel are at a heightened risk of infection since they are constantly exposed to the virus on a daily basis. This makes the lack of PPE a dangerous situation for both the healthcare provider, their families, as well as others in the hospitals. Based on an estimate by the World Health Organization, it is predicted that approximately 89 million medical masks are required monthly amidst the COVID-19 response. In terms of gloves, 76 million are required each month and goggles are demanded at 1.6 million per month. Breakages within the supply chain and surging prices have limited the supplies of PPE in many regions and has contributed to its existing shortage (Chaib, 2020).

To combat this, communities across the globe have been utilizing other, typically non-medical technological devices, to help lend a hand in the battle against PPE shortages. One of these initiatives includes the crowdsourcing of 3D printing masks for healthcare personnel on the frontlines. As heightened demands have made it harder for hospitals to receive PPE from formal suppliers, certain individuals have made 3D printing files for masks, goggles and other essential PPE available on the internet so that those with access to 3D printers can help bridge the gap in the supply chain (The Top 5 Practical Digital Health Technologies in the Fight Against COVID-19, 2020).

**Technological Demand (Other Affected Parties)**

While COVID-19 patients have recently been the priority of many healthcare institutions, it is important to not overlook other patients that require help unrelated to the virus. Unfortunately, the pandemic has undoubtedly impacted the medical routines and needs of many individuals and has resulted in detrimental effects on these patients and their families. When interviewed, Dr. Bruce K. Lowell, an internist in Great Neck, New York, mentioned that the virus is depriving other patients of desperately needed treatment and can result in deaths for non-COVID-19 related reasons (Grady, 2020).

As aforementioned, ICU beds and resources are all being gathered in efforts to fight the COVID-19 pandemic, however, it places these resources in short supply for patients who are not infected by the novel virus. Operating rooms have been turned into ICU rooms and surgeons and other specialized medical staff have been redirected to other departments to treat patients who suffer from the COVID-19 virus. Due to these changes, approximately one in four cancer patients have reported delays in their care due to the pandemic, causing adverse effects on their recovery and even resulting in death (Survey: COVID-19 Affecting Patients' Access to Cancer Care, 2020). Organ transplant surgeries are also heavily impacted since the conversion of operating rooms into ICU facilities after hospitals have been overwhelmed limit the ability to remove organs from donors and perform transplants. In the Columbia New York-Presbyterian Medical Centre, transplant surgeon Dr. Kasi McCune notes that the average number of transplant surgeries occurring each week has dropped from 750 to 350 every since the pandemic began in New York (Grady, 2020). While there is a shortage of technology to treat patients infected with COVID-19, patients with other serious medical conditions are also not receiving the care that they need and become non-virus-related victims of the pandemic.

Aside from the more technologically demanding health conditions that require surgeries or ventilator support, many individuals are also hesitant to visit hospitals out of fear that they will contract the virus. Even if there are genuine medical concerns, patients will often opt to stay home instead of seeking medical attention. A new program that certain hospitals have adopted is the integration of "telemedicine" into healthcare services. Through telemedicine, healthcare professionals can consult patients either through voice or video calls and provide contactless care from a distance. Primarily used for less critical medical concerns, telemedicine has taken full advantage of modern technology to help ease the current technological demands of the healthcare system (Breen & Matusitz, 2010).

## Disparities in Access to Technology

Demands for increased quantities of health technology have been rapidly rising since the start of the pandemic. While large urban hospitals are struggling to manage the large influx of new patients each day, those in rural regions are inadequately equipped to combat a pandemic outbreak. Specifically, with regards to the United States, rural areas in America have a lack of ICU availability in addition to a significant disparity between the distribution of healthcare resources (such as workers and funding) in comparison to urban areas (Ricketts, 2000). Especially in times such as a pandemic when urban hospital centres are overwhelmed with their own set of patients, rural hospitals and healthcare centres are often left on their own.

Residents of rural regions have fewer sources of healthcare in comparison to those in large metropolitan areas and have fewer healthcare professionals and hospital beds per capita (Erwin et al., 2020). Rural communities tend to have an older population and a higher rate of obesity, both of which are factors associated with an increased risk of mortality from the COVID-19 virus (Bialek et al., 2020; Dietz & Santos-Burgoa, 2020; Trivedi et al., 2015; Wong et al., 2019). However, a large contributor to morbidity and mortality surrounding the virus is the shortage of ICU facilities with ventilator support that is available to rural areas. These rural regions have also historically experienced the detrimental effects of limited health resources as seen in high H1N1 mortality rates in rural Turkey. High mortality rates in rural Turkey were directly linked to the deficiency of advanced ICU facilities and a shortage of ventilators (Kirakli et al., 2011). Additionally, individuals living in rural regions that are less developed or less wealthy may also have limited access to luxuries such as laptops, phones, or stable internet connection which are all necessities required for telemedicine. This lack of technological resources, both in the hospital and in homes, has the potential to hinder local pandemic recovery (Peters, 2020).

Other communities that experience disparities in access to health technologies include the Indigenous community and homeless communities. Both populations are already struggling with barriers to health including but not limited to access to clean water, housing, and poor mental health (Mercurio, 2020; Perri et al., 2020). Indigenous communities are faced with poverty and poor socio-economic conditions that limit their access to technological resources needed in order to help fight the pandemic (Mercurio, 2020). Furthermore, homeless individuals are also at an increased risk of infection due to the lack of safe housing and are also limited in their access to potentially life-saving technology (Perri et al., 2020).

The demand for technology extends beyond that of urban epicentres of the pandemic. In ill-equipped regions such as rural areas, Indigenous communities, and homeless communities, many individuals do not have access to the same devices and tools provided in urban centres. While there is still an increasing need for technology in these communities, the healthcare systems in rural regions, Indigenous reserves, or in the homeless community are unable to supply and meet this demand.

# Chapter 12: Dependence on Technology

As COVID-19 continues to dominate our lives, one thing that has come to the rescue is technology. It has been used as a tool to attempt to flatten the curve, help distribute supplies, and allow people to communicate in a world of social distancing. Society has fallen victim to a silent enemy that has been able to save us in many ways, but is equally capable of being an unwanted force in our world. From medical technologies to systems that have allowed workplaces to continue functioning, it is evident that COVID-19 has resulted in an increased dependence on technology. It is unclear whether this shift is for better or worse, but nevertheless, it has affected each and every life in countless ways. Schools and universities, having already made use of technology on campus in so many ways, must now take their skills to another level to offer high quality education online. Children in kindergarten and preschool may become part of a generation that only knows how to communicate through technology. With regards to healthcare, although PPE was always standardized and required, workplaces that were not typically stringent about hazards are now scrambling to purchase extra protective equipment. Finally, many companies have developed surveillance technologies to determine who has COVID-19 in order to alert those around them. This has resulted in a surge in technological innovations, thus increasing our dependence on technology. These aspects of this 'new normal' will be explored in the following chapter (Hendry, 2020).

## Changes in the Workplace

When the COVID-19 pandemic struck in mid-March, one of the first effects that was immediately seen was a shift in the structure of workplaces. Some workplaces are inherently riskier than others. Office settings are generally lower risk as there is a greater potential for physical distancing. Some of these businesses simply shut down, while others attempted to operate from home. This dilemma that most businesses had to face called for a modern workforce management solution. Normally, these solutions would involve ways to create a better group dynamic among employees, but during a pandemic, technological solutions are employed to ensure workplace safety and business continuity (Hendry, 2020). Technology is used in the form of a pre-existing automated workforce management system to reach out to employees with a pre-shift wellness questionnaire. There is also an increased importance of touchless options for a variety of services. For example, mobile devices are now being used for employees to clock in and out of shifts. Temperature taking technology is also on the rise and is necessary in workplaces to control the risk of COVID-19. Workplaces that seem to be riskier include meat-processing facilities and other facilities in the industrial sectors. Some of the largest outbreaks in Canada have been traced back to meat-processing facilities. Companies such as Blackline Safety have been working on solutions to track the movement of workers on a job site in efforts to ensure social distancing requirements are being met. This technology also allows for contact tracing and identification of works who may have come in contact with the virus (Bakx, 2020). Since many workers may not be able to carry a smartphone while on the job, Blackline is also developing a wearable device with the same functions. In addition, the National Health Service in Britain is developing an app that uses Bluetooth signals to log when smartphone users are within close proximity to one another. If a person develops COVID-19-like symptoms, an alert will be sent to other users of the app. This low-tech model is likely to be more effective and usable since it does not raise concerns over data privacy. The app is also anonymized so that users will not be told which person triggered a warning (Bakx, 2020).

Another way in which workplaces have utilized technology is through video monitoring. Amazon has launched an AI-powered system called 'Distance Assistance' that provides feedback to employees on social distancing (Mathur, 2020). The program has a monitor displaying a live video with indicators to show if employees are 6 feet apart from one another. Another company called Staqu, which is an Indian startup, is also using video analytics to monitor employees to identify and track the spread of COVID. Video analysis is a cost-effective solution that allows employees to return to office settings safely.

While maintaining physical distance is important to prevent the spread of COVID-19, companies have also had to develop cleaning technologies. Floor-disinfecting robots have been utilized to speed up the cleaning process. This technology was developed by Milagrow Humantech and can navigate and sanitize floors with no human interventions. It destroys COVID on surface using a sodium hypochlorite solution. This particular robot uses LIDAR and SLAM technology to scan surrounding environments and rapidly develop a floor map. MIT has also designed a floor disinfecting robot in partnership with Ava Robotics that uses a custom UV-C light to disinfect surfaces and neutralize aerosolized forms of SARS-CoV-2. UV-C light have proven to be effective at killing viruses on surface, but it can be harmful to humans. Thus, the robot does not require any human supervision (Kim, 2020). Air purification systems have also been developed by a startup called Magneto CleanTech. This system also uses UV-C beams to sterilize indoor air and can kill over 90% of airborne infection (Mathur, 2020). The standard air filter used in the US is HEPA, which stands for high-efficiency particulate air. It removes 99.97% of particles in the air. Thus, aerosol droplets can be filtered out with this system. It will not stop the spread if a person is living with someone who is contagious seeing as the air purifier will not have the chance to remove particles in such a short period of time (Priest, 2020).

Some other ways in which technology has assisted during this pandemic are through platforms such as Zoom which have facilitated remote work. Companies have survived using Zoom, Webex and Microsoft Teams. It is likely that there will be a rise in the development of more communication technologies. There has also been an increase in voice control devices to reduce the frequency of contact points to promote safety. Some technologies have allowed users to dictate and send emails. It can also be used to make calls, schedule appointment and transcribe meetings. Furthermore, AI has been used in Human Resources. It can now be used to write job discrimination, allow job candidates to schedule their own interviews, and ensure fairness and anti-discrimination. AI is also able to evaluate employee performance by gathering data analytics.

Drones are another advanced technology that have made the markets during this pandemic. They have been used to efficiently scan areas and broadcast messages telling people to wear a mask. This also allows for officials to distance from potentially infected people. Drones have additionally been used to deliver critical supplies. The pandemic has restricted package and food delivery systems as they may be contaminated, but drones provide a contactless alternative to provide medical supplies (Priest, 2020).

While contact tracing is a technology being utilized in the workplace, it is also being developed on a larger scale. Ontario is working to use a contact tracing app to prepare for the second wave of COVID-19. This app uses Bluetooth technology that notifies users if they have come in contact with a person who has tested positive for COVID-19. It also employs the use of Global Positioning Systems. While this concept seems solid in theory, there are concerns about the lack of evidence pertaining to the practicality and accuracy of such apps. Some have said that this initiative should be backed up by human checkers to ensure accuracy.

**Changes in the Education System**

COVID-19 has drastically changed the scene when it comes to education. Students were excited for an extended March Break, but little did they know online school would soon become the new normal. There are approximately 1.2 billion children in 186 countries who are now out of the classrooms. Many tech companies are doing their part in making online education more accessible. For example, Google has recently launched Teach from Home which is a hub that offers support for teachers to improve remote learning. Microsoft Teams, Zoom and Slack are all being used to facilitate online learning. Further, BYJU, which is a Bangalore-based educational technology firm, is announcing free live classes on its Think and Learn App. This company is now the most valuable edtech Indian startup, with about 7.5 million new users since it began offering free access to content (Mitter, 2020). The live classes offered on this app aim to replicate a classroom environment by having periodic scheduled classes, thus increasing productivity of students.

Another platform called Tencent Classroom, a China-base company, has allowed for students in China from K-12 to attend online school.

Some may believe that the unplanned switch to online learning may result in poor quality of education, while others believe that a new hybrid model of education will emerge with significant benefits. Online learning allows teachers to reach out to students more efficiently, and platforms such as Lark make it easier to communicate. Lark is a Singapore-based collaboration suite developed as an internal tool, but now offers teachers and students unlimited video conferencing time. Lark has ramped up its server infrastructure to ensure reliability. Some teachers have even said that they will continue using such platforms even once the pandemic has settled down. At the same time, there are obvious challenges to digital learning. For example, those without reliable internet access or high quality may not be able to participate in online lectures. For those who do have the right technology, it has been shown that online learning can be more effective in several ways. It may improve memory retention during class since it is possible to learn faster. The education system has a general consensus, however, that structured classroom education is vital to a child's development because children are generally more distracted online. For online learning to truly be useful, there must be an effort towards making the classes structured to promote inclusion, personalization and intelligence (Li, 2020). Prior to the pandemic, many parents leaned towards having their children participate in online classes because it provides them with more control over what their kids are learning. There is a lack of research on how online school may affect children's socio-emotional skills (O'Hanlan, 2020). Overall, the new education system is a prime example of how COVID-19 has increased our dependence on technology. Many teachers have reported that the pandemic has improved their views and ability to use technology as they have had to quickly transition to a new model. Teachers have evidently struggled more than students with this transition and individuals in younger generations have been largely raised in an environment with technology.

The COVID-19 pandemic has led to technology being at the forefront of our lives now more than ever before. Workplaces have had to develop new and innovative solutions to promote safety within offices and industrial settings, while also increasing productivity in an online setting. Robots and drones are just some examples of artificial intelligence-based technologies that have improved contactless cleaning and delivery of products. As described in previous chapters, our world is currently dependent on medical technologies that will hopefully result in the early development of a COVID-19 vaccine. Children around the world are now dependent on technology in different ways than before. While technology may have been an outlet to communicate with friends and play games, it is now their vehicle of education. Elementary schools, high schools and universities have all had to invest in technological platforms that will allow students to have the maximum educational benefits. It is clear that each and every person has an increased dependence on technology in some way or the other. The question is, what are the long-term impacts of having our lives centered around the use of technology?

# Chapter 13: Economic Impact of COVID-19 on the Technology Sector

## Introduction

The COVID-19 pandemic has caused drastic changes to the way the world operates. Many countries have had to close down borders and shut down factories to limit the virus's spread. Changes have also taken place in consumer behavior that could drastically shift many of the behaviors that have been a part of the norm for many years. This brings with it many positive and negative impacts in both the short and the long term which will be explored in this chapter.

## Negative Impacts

The COVID-19 pandemic is being described as a "black swan event": a rare, unpredictable event that has potentially severe consequences on everyday life (Deloitte, 2020a). These effects have already been felt in the technology industry in the form of supply chain disruptions, decreases in demand and delayed product launches, and will continue to persist in the future as well (Deloitte, 2020a).

*Short Term*
In the short term, social distancing measures have caused supply chain disruptions, decreasing demand and a delay in the release of new products that have led to significant revenue and profit loss for many technology companies. In fact, "tech's big five[Apple, Microsoft, Amazon, Alphabet and Facebook] [had lost] $1.3 trillion in value" by mid-March of 2020 (Levy, 2020).

## Supply Chain Disruptions

As the economy becomes increasingly globalized, many tech companies have expanded their supply chains to include geographically distant locations from their main base of operations. This was done mainly to take advantage of benefits like "low-cost labor, favorable tax structures, and synergies with both suppliers and customers". While these benefits are certainly advantageous for the company to maximize revenue, it has created a risk of geographical concentration of the supply chain, where companies are overly reliant on one country or region for much of their raw materials and production capabilities, which is called a single point of failure. This makes their supply chain less resilient when faced with unexpected circumstances that severely compromise that region (Deloitte, 2020a). This has certainly been the case during the COVID-19 pandemic, during which many manufacturing hubs like China and other East Asian countries have faced travel restrictions and lockdown (Langton, 2020). These disruptions have caused a "shortage of components" and "[created] choke points" in their supply chains (Deloitte, 2020b). In fact, according to a study done by Dimensional Research, 37% of tech companies surveyed said they are unable to fulfill customer requests (Supply Chain Quarterly, 2020). This led "94% of the Fortune 1000, including many technological companies, [to suffer from] supply chain disruption" and has contributed to the losses in revenue and negative economic impact previously described (Sherman, 2020).

## Decrease in demand for certain tech products

While supply has certainly been affected by the COVID-19 pandemic, consumer demand for certain tech products has also fallen. This has occured because, as many workers lose their jobs and income, they also reduce their spending on anything nonessential (Andersen, Hansen, Johannesen, & Sheridan, 2020). Many of these tech products are a discretionary purchase, and replacing them regularly is not a necessity. One industry that has been severely impacted is the smartphone industry. According to studies done by both Counterpoint Research and Canalys, global shipments of smartphones have dropped 13 percent in the first quarter compared to the previous year (Byford, 2020). Another sub-sector that could be significantly impacted by falling demand due to

COVID-19 is television, with major brands like Samsung and LG reporting that TV profits are expected to "decline significantly" (Porter & Ricker, 2020). This fall in demand has significant negative consequences for companies in the tech industry in the short term including loss of revenue, layoffs and a slowdown of innovation, which may persist until the impact of the virus has been managed.

**Delay in the release of new products**

Due to disruptions in the supply chain and waning demand caused by COVID-19, many companies in the tech industry have been forced to delay or cancel the launch of new products. In fact, according to a Dimensional Research study, "53% of electronics industry product launches have been delayed or cancelled due to the pandemic" (Jasinski, 2020). This includes industry giants like Apple, which could reportedly delay the launch of its next iPhone by "months"(Haselton, 2020) and Microsoft, which is delaying the launch of its Surface Duo and Neo devices (Walsh, 2020). This negatively impacts the tech industry as a whole and has contributed towards the consequences previously described.

*Long Term*

**Worsening divides that already exist in the technology sector**
Many of the big, well-capitalized tech companies have benefited from the COVID-19 pandemic. In fact, the "tech-heavy NASDAQ-100 Index has risen more than 40 percent" since March, led by the FAAMG stocks of Facebook, Amazon, Apple, Microsoft and Google (Alang, 2020). This stems not only from the usefulness of these products as people's behaviours shift during the pandemic, but also because these larger companies are most prepared to quickly adapt to the new norm (Wakabayashi, Nicas, Lohr, & Isaac, 2020). In contrast, smaller tech companies have struggled during the pandemic due to supply chain disruptions and difficulty finding capital, which has been reflected through decreased revenue and lower share prices (McCaffrey, Moline, & Palladino, 2020). This has caused these larger, well-capitalized tech companies to "[expand] their power by grabbing talent and buying companies for their [Intellectual Property] - then dissolving them"

(Saade, 2020). This is expected to increase over the long term as the effects of the pandemic are fully realized. By creating an environment where big tech companies become stronger and smaller companies become weaker, the pandemic worsens the pre-existing divide between the two and increases the likelihood for monopolistic behavior by Big Tech in the future (Saade, 2020).

**Slowdown in innovation and discovery of new opportunities for partnership**
The effects of the COVID-19 pandemic can also slow down innovations and the discovery of new partnership opportunities in the tech industry, which can negatively impact firms' economic outlook in the future. This is first because of industry event cancellations. These events, like E3 and SXSW, give members of the tech community the opportunity to connect with one another and share ideas, which is crucial towards driving innovation and developing new opportunities for partnership. The cancellation of these events can thus contribute to a slowdown in innovation and the development of new partnerships in the future (McCaffrey et al., 2020). The slowdown in innovation may also occur because of a decrease in the recruitment of skilled workers as tech companies cut costs during the pandemic (McCaffrey et al., 2020). A constant expansion of the workforce to include new skilled workers and new ideas is crucial to the culture of innovation created at many of these tech companies (Chandler, Keller, & Lyon, 2000). However, due to the pandemic, many companies have slowed hiring to cut costs as revenues decrease (Reed, 2020). This can affect the development of new innovations in the future, which can negatively impact the industry.

**Positive Impacts**

The COVID-19 pandemic has the unique opportunity to serve as the catalyst for an even closer relationship between humanity and technology as individuals and businesses around the world adapt to the new norm that has been created. This has significant short term positive effects on the technology sector, defined as the time period when social distancing measures remain in place. This will also carry over into the long-term, after the threat of the pandemic has mostly passed, as the long term effects of the virus are felt and tech companies strategize to prevent similar outcomes in the future.

*Short Term*
One of the biggest benefits of the pandemic for the tech industry is the change in people's behaviors in an effort to slow the spread of COVID-19. This includes a rise in remote work and online education, e-commerce and social distancing that have caused increased demand for products from technology companies that facilitate these outcomes in the short term.

**Remote Work and Online Education**

Some of the biggest beneficiaries of the lockdown procedures enacted around the world are companies that facilitate the shift towards remote work and online education. This includes tools that are essential to replicate the work and school environment, such as remote desktop applications like Teamviewer, cloud storage options like G Suite and mobile hotspot software like Karma (Pinola, 2020). This also includes virtual collaboration & communication tools, like Slack, Zoom, Skype and Microsoft Teams, which allow members of institutions to seamlessly communicate with one another while working from home (Pinola, 2020). In March of 2020, as the pandemic spread around the world, video-conferencing apps "saw a record 62 million downloads"; an indication of the boost in demand that such virtual communication tools have gained (Wood, 2020).

Another significant sub-sector that stands to gain from the transition to remote work is cybersecurity. According to Larry Zelvin, Head of Financial Crimes at BMO Financial Group, since March of 2020, "criminal fraud activity has increased by up to 800%, with a number of attacks taking place against pharmaceutical operations, hospitals, and even the World Health Organization" (Amle, 2020). In fact, according to the Absolute 2019 Global Endpoint Security Trend Report, "42% of endpoints [which refers to any remote computing device that communicates with a network, such as Desktops and laptops (Smartbear, n.d.)], are unprotected at any given time" (Ahmad, 2020). These statistics highlight the need for increased cybersecurity as much of the population shifts to working from home. This will especially be prominent in the cloud security market, as it is predicted to grow 33.3% this year. Furthermore, data security, application security, and identity and access management markets will also see growth, predicted to be anywhere between 5% and 10% (Scroxton, 2020). This has led to increased demand for companies like Okta, which is used to control and monitor network access, and Zscaler, which is used to decongest VPN traffic (Ahmad, 2020).

**Online Shopping**

As lockdown procedures around the world have restricted people from frequenting malls and grocery stores, many have turned to online shopping to meet their shopping needs. In fact, as of April 21, there has been a 129% year-over-year revenue growth in Canadian and U.S. e-commerce orders; an indication of a drastic boost in online consumer spending during the pandemic (Emarysys, 2020). This staggering increase in online retail has partially been fuelled by an increase in impulse purchases, with many consumers using retail therapy to cope with the mental stress of the pandemic. This increase is also partially the result of an increase in panic buying, as people purchase in excess in fear of running out of supplies in the future (Italie, 2020). Both of these psychological effects are described in detail in Chapter 10. While a large fraction of online retail businesses have benefited from the pandemic, companies producing certain, highly-sought after products have done especially well. This includes suppliers of health products, groceries, canned goods and fitness equipment.

*Long Term*

**An increased focus on de-risking the supply chain**
As described earlier, the supply chains of many companies have become overly reliant on one country or region for much of their raw materials and production capabilities, which has created a risk of geographical concentration of the supply chain (Deloitte, 2020a). The COVID-19 pandemic has demonstrated the importance of this issue and has caused many companies to begin focussing on reducing supply chain risk and the likelihood of disruption in the future (Zanni, 2020). One strategy for this is a diversification in the location of production facilities so that, if

a disruption occurs in one region, plants in other regions can compensate to reduce supply shortages. Wistron Corp, a supplier for Apple, is one of the companies looking to do so, having indicated that half of its capacity could be located outside China by 2021(Wu, 2020). Another strategy would be to build up a supply of inventory safety stock that would help meet demand in the case of a disruption to supply (Binder, 2020). Companies are also considering having suppliers closer to consumption markets so that supply and demand in that region are affected equally and a surplus or shortage will not come to exist (Binder, 2020). This increased focus on de-risking the supply chain can help prevent supply disruptions in the future, which is beneficial for technology companies in the long run.

## Establishing permanent behaviors

While many shifts in peoples' behavior are a direct response to the pandemic, they may also result in lasting change to how society functions. Historically, crises like the COVID-19 pandemic have caused a modification to social norms and behaviors, such as the shift towards increase airport and building security after 9/11 (Martinez, 2020). The COVID-19 pandemic could similarly cause lasting changes in peoples' behavior especially in relation to technology. One sub-sector that could benefit from this is online shopping. Even after the pandemic has passed, consumers may try to avoid physical stores and instead resort to online shopping, which is not only safer, but more convenient . This is especially the case with the many older, first-time users during the pandemic who have resorted to online shopping because of the "disproportionate" impact that the virus has on members of this age group (Martinez, 2020). This has opened up e-retailers to an entirely new demographic which could help drive sales in the future.

# Chapter 14: Technology as a Driving Force for Political Changes

## Introduction

As demonstrated throughout this publication, technology has become more prominent during the COVID-19 pandemic, as many aspects of modern life are moved to virtual platforms. This has caused an acceleration in tech adoption that has left governments around the world scrambling to institute measures to promote the fair, equitable usage of technology (Baig, Hall, Jenkins, Lamarre, & McCarthy, 2020). Furthermore, this technology has also made governments privy to new forms of attack during the pandemic for which many are not prepared for (WHO, 2020e). On the other hand, tech has also helped keep the government functional since it has provided a way for members to stay in contact with one another, either via call or video (Tasker, 2020). Moreover, it has also provided a platform for governments to spread important information to a large number of people, helping ensure public health and safety (United Nations, 2020). These impacts of technology on politics and government during the COVID-19 pandemic will be evaluated in this chapter.

## Difficulty ensuring fairness and equity in the tech industry

Ensuring fair and equitable usage of technology has caused struggles for governments all around the world for many years. Officials have constantly grappled with both intervening too much, and gaining the reputation of a surveillance state, and intervening too little, and enabling uncontrolled private sector behaviour (Milano, 2019). These effects have only been heightened during the COVID-19 pandemic, demonstrating the importance for an effective government approach to the issue.

*Predatory Mergers*

An example of this is the increase in predatory mergers during the pandemic (Lewis, 2020). Many investors see the chaos caused by the COVID-19 outbreak as a rare chance to purchase assets at discounted prices and have thus engaged in increased asset acquisition. As Senator Elizabeth Warren explained, "as [society] fights to save livelihoods during the [COVID-19] pandemic, giant corporations and private equity vultures are waiting for a chance to gobble up struggling small businesses and increase their power through predatory mergers" (Lewis, 2020). This poses a serious risk for the country and the economy, since many of these small businesses are subsequently broken up and sold for parts, taking away the numerous jobs that they provided as well (Saade, 2020). To combat this, many are working to curb this predatory behavior, at least until the pandemic ends. Two of the biggest voices on this issue are Senator Elizabeth Warren (D - Mass.) and Rep. Alexandria Ocasio-Cortez (D-N.Y.), who have worked together to introduce legislation to halt large acquisitions and mergers during the COVID-19 pandemic (Lewis, 2020). This act, dubbed the Pandemic Anti-Monopoly Act, would institute a moratorium on takeovers involving companies with greater than $100 million in revenue and financial institutions with a market cap over $100 million. However, this bill is a long shot, as it would require approval from the senate, which is currently under Republican control (Lewis, 2020). Furthermore, on July 29th 2020, CEO's from Tech's 4 biggest firms, Amazon, Apple, Facebook and Google, were brought in front of lawmakers to testify on their companies alleged antitrust and monopolistic behavior as part of an investigation into the matter that has been ongoing since June 2019 (M. Kelly, 2020). This represents a significant step in curbing monopolistic behavior in the tech industry. While such actions are being taken, more needs to be done in the short term to protect small businesses struggling during the pandemic from predatory behavior.

*Misinformation*

Another issue politicians are grappling with during the pandemic is misinformation being spread through technological platforms. Misinformation refers to false/inaccurate information that is deliberately intended to deceive individuals (Desai, Mooney, & Oehrli, 2020) This has become increasingly prevalent during the COVID-19 pandemic, with 10% of adults and 31% of children and teenagers having shared fake news stories online, a study done in May of 2020 found (Watson, 2020). Some of the topics that are commonly reported on incorrectly are the origin of the virus, its spread, and potential treatments (Ball & Maxmen, 2020). This misinformation can have devastating impacts. For example, after President Donald Trump tweeted that the drug Hydroxychloroquine was a "cure" for COVID-19, it was found that this drug instead led to an increase in deaths (Boseley, 2020). Similarly, when Biohackinfo.com incorrectly claimed that Bill Gates planned to use the COVID-19 vaccine to monitor individuals, it gained two million views on youtube, with Roger Stone, who is a former adviser to President Donald Trump, discussing the theory on a radio show, resulting in even more views (Ball & Maxmen, 2020). According to Joan Donovan, a sociologist at Harvard University, this post had "better performance than most mainstream media news stories". When a vaccine does emerge, this might discourage many individuals from taking it, which can lead to the impacts of the virus to continue for many years, thus making combating misinformation a significant issue for governments to tackle (Ball & Maxmen, 2020).

To this end, many governments around the world are implementing measures to combat the spread of misinformation. Countries like Singapore, Germany, Malaysia, France and Russia have already passed laws that would make it an offense to spread misinformation knowingly while Canada and the United Kingdom are discussing implementing such measures (Schetzer, 2019; Thompson, 2020). However, this brings with it many concerns Through this approach, government ministers are given the right to determine what is considered fake news as well as the authority to demand online platforms to remove content that goes against their policies. This has alarmed human rights activists, legal experts and others who believe that these laws can be used to restrict free speech or unintentionally block legitimate sources of information (Schetzer, 2019).

## Pandemic-related Cyber-Warfare and Theft

Another contentious issue affecting many governments around the world is the increase in Cyber-Warfare with respect to COVID-19 related technologies. Companies and institutions in countries around the world are racing to develop a vaccine and other treatments for COVID-19, not only because it would significantly aid the recovery of its home country, but because it would bring in a significant amount of revenue as much of the world awaits in anticipation (Cohen, 2020). Furthermore, many countries are fearful that the distribution of the vaccine would be guided by world politics, where the nation responsible for the discovery would share the vaccine to its close allies, before providing it to other countries, if at all (Kupferschmidt, 2020). This has led to an increase in cyber-warfare and theft against corporations and institutions working to develop COVID-19 vaccines and treatments (WHO, 2020e).

On July 16th 2020, the American, British and Canadian governments accused the Kremlin of attempting to steal COVID-19 vaccine research (Barnes, 2020). This was specifically targeted at the Russian intelligence group known as Cozy Bear, which was implicated in the 2016 break-ins of the Democratic Party servers. The group was accused of trying to steal vaccine and treatment intelligence from universities, corporations and other affiliated organizations and of attempting to "exploit the chaos created by the [pandemic]" according to officials. This was mainly done using malware and fraudulent emails that attempt to trick employees into turning over security credentials in order to gain access to vaccine research and information about medical supply chains. One of the most frequently-targeted institutions is Oxford University in Britain and its partner, the British-Swedish pharmaceutical company AstraZeneca, who have been working jointly on the vaccine. Oxford scientists reported in July of 2020 that they noticed a "surprising resemblance" between their vaccine and the work of Russian scientists, both attempting to alter a common virus to mimic COVID-19 to cause an immune response without leading to illness (Barnes, 2020).

Other nations actively working to steal COVID-19 vaccine research in the US include China, North Korea and Iran. In July of 2020, two chinese nationals, Li Xiaoyu and Dong Jiazhi, were indicted by the United States for attempting to steal COVID-19 vaccine research and for hacking hundreds of companies both in the US and abroad, according to the Justice Department (The Guardian, 2020). Similarly, hackers from Iran were caught attempting to gain access to Gilead Sciences, the manufacturer of Remdesivir which is a drug approved by the FDA for clinical trials (Sanger & Perlroth, 2020). They were also accused of attempting to cut off the supply of water to rural parts of Israel while many were confined to their homes. Even countries traditionally seen as allies of the west have been accused of engaging in fraudulent activity. One example is South Korea, which has been accused of attempting to hack into the World Health Organization as well as into the accounts of officials in the United States, Japan and North Korea (Sanger & Perlroth, 2020).

Protecting a nation and its corporations' cybersecurity during these times is a significant challenge for governments (T. Kelly, 2020). Many of the attacks taking place target individuals working in a specific organization with scam emails and posts that prompt the individual to provide their security credentials instead of attacking other entry-points that are traditionally well-protected (T. Kelly, 2020). The only way to mitigate the impact of this form of hack is to educate the overall workforce and put identity verification measures into place (T. Kelly, 2020). Furthermore, governments also struggle to identify the location from which a hack originated and the nation responsible, since many hackers are now able to make it appear as though the hack originated from a completely different location (Hackett, 2016). This ambiguity makes cybersecurity and theft a controversial yet paramount political issue that governments must address both now and in the future.

**Governments usage of technology for communication**

Technology has also played a vital role in government communication during the COVID-19 pandemic. One of the biggest uses for it has been for members of government to communicate with one another as well as with allies abroad. In April of 2020, Canada's House of Commons met for the first time in a virtual setting using the platform Zoom (Tasker, 2020). While available to the public en masse, the House of Commons used a reconfigured version of the platform that had security features different from the paid and free consumer versions. Furthermore, unlike other meetings, parliamentary proceedings are open to the public, and thus there was no risk that confidential material is leaked. However, the platform has faced concerns about security that serves a barrier for increased usage (Tasker, 2020). In the past, Jewish users have seen religious services hijacked (Brown, 2020) and a Zoom video chat conducted by a black NHL player was hacked and inundated with racial slurs (The Guardian, 2020a). A study conducted by the University of Toronto's Citizen Lab found that Zoom did not use "true end-to-end encryption" and that the company has the "theoretical ability to decrypt and monitor Zoom Calls" (Tasker, 2020). Moreover, the researchers at the University of Toronto found that the Zoom encryption keys, which are a set of random letters and numbers used to scramble and unscramble transmitted data, were sometimes routed through servers in China even if the meeting's participants were not geographically located in China. Concerns like these have led to the publishers of the report at Citizen Lab to label Zoom and other such platforms as "a gold rush for cyber spies" and that "governments worried about espionage ...[should discourage the use of Zoom]" (Tasker, 2020). Thus, while technology can help maintain a sense of normalcy in government procedure during the pandemic, security and safety concerns must first be identified and corrected.

Furthermore, technology has also been used by the government to communicate effectively with the public. Especially during times of crisis, it is important for the government to provide up-to-date, accurate and useful information to its citizens (United Nations, 2020). According to a review of the national portals of the 193 UN Member States, by the 8th of April 2020, 86% of countries had included information and guidance about COVID-19 on their portals; up from 57% in March (United Nations, 2020). This comes in a variety of ways. The most basic form on some national portals is a re-posting of media coverage that informs citizens about the outbreak, governmental response, protection procedures and travel restrictions. Some governments have also sought a more advanced method, with a dedicated section about the outbreak with statistics and location-specific information. Providing this reliable information is crucial since it allows citizens to make informed decisions and promotes public health and safety (United Nations, 2020). Governments have also used technology during the pandemic to engage people, especially those that come from socio-economically challenged backgrounds, so that no one is left at a disadvantage. One example of this is Government-organized hackathons where public officials, along with software developers and social entrepreneurs, work to find new, innovative ways to tackle technological, social and economic challenges brought on by the pandemic (United Nations, 2020). This was seen in China, where through such an initiative, local governments launched a city Health QR code service that gathered health data declared by all residents (United Nations, 2020). Governments have also partnered with influencers on social media to counter misinformation and spread accurate information about the pandemic, especially amongst youth, who are most vulnerable to fake news spread online (United Nations, 2020).

# Chapter 15: The Long-Term Impacts of COVID-19

Although Covid-19 will eventually be beaten and the world will exit out of the pandemic, a global event of this scale will have long-lasting if not permanent, consequences on almost every aspect of everyday life. It will bring changes to the workplace, economy, our culture, how we prepare for future pandemics, and so much more, and much like pandemics and other major global events of the past, it will leave our world in a new state and force us to adjust to a new normal. Although it is impossible to know for certain what the future after the pandemic will look like, there are numerous predictions that can be reasonably made based on information that is currently available.

## Future Changes

*Workplace*
The pandemic has forced many businesses to switch to a remote mode of operation, with teleconferencing and remote working now being commonplace as many digital companies such as Zoom have had the opportunity to optimize their software to best support a virtually connected workplace. With some workers still being able to maintain the same level of productivity from home compared to the workplace, there may be a reduction to the in-person work week and more people simply working from home altogether, which would have many benefits for both the employers and the employees. Changing business models to a 4 day in-person work week, for example, would reduce operating costs for employers as they now have to only keep the office open for 4 days per week, while also reducing expenses such as gas for employees as they no longer have to commute as much (Benstead, 2019). The digitization of work reducing mobility needs among the general population will also have the added benefit of reducing fossil fuel emissions, which would be a step in the right direction towards fighting

climate change (Kanda & Kivimaa, 2020). Although this may, of course, not be possible for every business model, an increase in the number of businesses switching to a model of a reduced in-person workweek will have significant benefits overall.

*Economy*

COVID-19 has devastated and changed countless aspects of the global economy. Many countries saw a sharp drop in economic output as businesses were forced to close and supply chains were compromised. This led to a significant decrease in the Real GDP of many nations, with the International Monetary Fund predicting a global fall in Real GDP of 3% (Gopinath, 2020). In fact, it is estimated that the cumulative output loss from 2020 and 2021 could be as high as $9 trillion, making this the worst economic downturn since the Great Depression (Gopinath, 2020). As a result of necessary actions taken by governments, such as stimulus bills to reduce the economic decline, it is no surprise that debt levels have also risen. Debt is estimated to rise in some economies, particularly those of developing nations, by as much as 5 percentage points of GDP, although the true impact on individual economies is determined largely by domestic policies and choices (Kose et al., 2020). The impacts of this are widely varying and complex but will impact the world years after the end of the pandemic before economies are fully able to recover.

As beneficial as globalization has been to the global economy and to both consumers and producers as a whole, it unfortunately has also been one of the greatest sources of weakness for companies amid COVID-19. Over the last several decades, the world has become increasingly interconnected and has thus built supply chains that were fast and efficient. However, recent events have led to these supply chains being recognized as "fragile and unstable"(Crabtree, 2020). This is because current global organizational structures were designed and built in a period that had considerably less complexity, but the rapid rate of technological development has left governments and corporations scrambling to keep up. The result of this is supply chains becoming more efficient than ever before due to the advanced technology that is being employed, but also less robust as they have become increasingly dependent on technology which brings about a far greater level of complexity (Fetterly, 2020). Due to the impacts of the pandemic, multinational corporations will most likely begin to focus on decreasing vulnerabilities and the simplification of supply chains, while nations will most likely be inclined to increase domestic production of certain goods to reduce their dependence on the global economy (Fetterly, 2020). This may, in turn, result in higher prices for some goods, as costs of production in countries such as Canada and The United States are higher due to higher labour costs and more regulations for businesses. In addition to these changes, governments and corporations alike will need to structure themselves for rapid adaptation to changes in the external environment. They must adapt an anticipatory approach to developing and incorporating new technologies and practices, and they must be willing to abandon whatever is already established (Fetterly, 2020). Overall, the aftermath of the COVID-19 pandemic will likely lead to businesses and governments placing more focus on reducing vulnerabilities and increasing reliability of supply chains and other necessary aspects, rather than focusing simply on efficiency.

*Travel*

Although certain flights are becoming once again, it will be a long time, potentially years, before the flying experience goes back to what it was prior to the pandemic. Many airlines worldwide have adopted new policies to ensure the safety of passengers which will remain in place for the foreseeable future. In addition to requiring every passenger to wear a mask, many airlines have implemented mandatory temperature checks prior to boarding (Jones R., 2020) and may subsequently deny boarding to passengers that present with a fever. Additionally, companies such as Air Canada are looking to take even more aggressive approaches to reduce the spread of COVID-19 by partnering with biotechnology firms to acquire rapid COVID-19 tests which will be administered to every passenger prior to boarding (Air Canada, 2020). With these measures soon to be implemented by Air Canada, it would not be surprising to see other airlines follow in its footsteps shortly after. According to the president of Emirates Airlines, there is some hope of air travel returning to normalcy by the summer of 2021, although this will depend on a vaccine being discovered by the first quarter of 2021 (Abbas, 2020). It is safe to say, however, that with the exception of COVID-19 rapid tests, some of these measures, such as temperature checks, may be left in place long after COVID-19 is no longer deemed a pandemic as a precautionary measure.

Unfortunately, many regions that depend on tourism as a large source of income have been some of the most hard hit in the world (Gopinath, 2020), but this may present an opportunity for those looking to travel once flights reopen. This is because many areas have begun offering incentives to tourists, such as reimbursement of plane ticket or other vacation costs, in an attempt to attract more tourists (Jones, D., 2020). This benefits both the traveller and the respective area as the traveller is able to visit certain destinations at a significantly cheaper cost while destinations are able to increase tourism levels which will hopefully assist them in restarting their industry.

*Culture*
Although there will surely be some long-lasting impacts with regards to culture and everyday life, these changes may be among some of the hardest to predict and will likely not be as significant as those in other areas. First, a sharp increase in divorce rates among couples is expected among couples, and this increase can already be seen. Toronto law firm Nussbaum Family Law said it is experiencing a 20 percent increase in inquiries from people looking to split from their spouses, with an expectation for this statistic to further increase after the pandemic. This trend is also already visible in other countries that are further ahead with regards to fighting the pandemic, such as China and Italy, who are reporting higher separation rates (Neustaeter, 2020). The way in which society views illness may also change, with individuals potentially beginning to take more precautions. It goes without saying that this pandemic has even devastated many who have not been infected with the virus, possibly leaving them with residual fear of becoming ill ever after the pandemic is gone. The result of this may be a cultural shift, with those who are feeling symptoms wearing a mask when in public to prevent the spread of whatever illness they are infected with. This would be similar to practices already adopted in parts of Asia long before COVID-19, as people will typically wear a mask if they are feeling sick. This is because it is seen as good hygiene and being considerate of others (Beglin, 2020), and the introduction of such a norm to western nations may reduce the spread of illnesses in the future.

## Sustainability Transitions

A key question for some researchers is how the current pandemic will influence currently unfolding sustainability transitions and what the general long term impacts on society will be, where a sustainability transition is defined as a shift from one socio-technical norm to another. Although these typically take several decades to complete, this pandemic has demonstrated that "some systemic and deeply structural changes in socio-economic systems can under certain circumstances occur quite rapidly." A key example of this would be the transition from private ownership of gas-powered vehicles to shared mobility solutions, such as ridesharing services and other transit options. Whereas many of these solutions have been on the rise over the course of the past decade, they are now seeing a decline in use as the worry over social distancing will lead to people favouring private vehicles over shared methods of transit (Kanda & Kivimaa, 2020). This is evidently seen through Uber stock, which from February 21, 2020 to March 20, 2020 dropped by close to 50% from 40.72 USD/share to just 21.33 USD/share. Although it has begun to recover since then as restrictions are loosened, their stock value as of the end of July 2020 was still approximately 25% lower than February 21, 2020. Overall, the pandemic has led to societies regarding public health as an absolute priority, and rightfully so, while significantly reducing focus on issues of sustainability. The result of this is a significant setback in certain sustainability transitions such as the one mentioned above.

## Preparation for Future Pandemics

Despite having experienced previous pandemics and having access to more advanced technology and an ever-growing knowledge base, it is safe to say that many of the world's nations were poorly prepared for the COVID-19 pandemic. Similarly to the rest of the world, the limited pandemic preparedness in Canada has led to costs incurred, both monetary and otherwise, far exceeding what would have been if governments were better prepared (Fetterly, 2020). However, the Covid-19 pandemic has without a doubt presented an opportunity for governments and society as a whole to learn from previous mistakes so that better preparation strategies and policies can be implemented for future pandemics

Achieving this will require significant planning and a change to how society views emergency preparedness. A potential first step would be to make emergency preparedness a federal priority and introduce annual reporting requirements; this can be accompanied by the creation of a department specifically for the purpose of emergency and pandemic preparedness. Additionally, investments in public infrastructure would greatly assist in the preparation of citizens for a changing environment and investments in healthcare systems to increase their capacity and capability in all aspects would be greatly beneficial in the event of another pandemic (Fetterly, 2020). Lastly, another possible step that may be taken is to increase funding and coordination in the scientific community. For example, the use of targeted funding programs to increase interaction between "medical scientists and more general biologists (natural scientists) from disciplines such as ornithology, ecology and evolutionary biology" (Christophersen & Haug, 2006) would enhance the understanding of the nature of evolutionary processes that affect influenza viruses (Fetterly, 2020). Many of these changes would also benefit the day to day life of citizens even when there is no threat of a pandemic, as infrastructure investments would increase the efficiency of economies while healthcare investments may lead to a better quality of care for people overall. The fact of the matter is that after a global event as devastating as the COVID-19 pandemic, it is crucial to learn from past mistakes and make changes at a high level to ensure that the world is better prepared to handle pandemics in the future.

Although reasonable predictions can be made based on the information presently available, it is, unfortunately, impossible to know what a future after the COVID-19 pandemic will look like as there are far too many factors to take into consideration, with one of the largest being when a vaccine will be developed. It is safe to say, however, that although life will eventually return to something fairly close to that it was prior to the pandemic, there will be some changes that will be made in a variety of fields. Overall, COVID-19 has upended normal life entirely and left its mark in human history. It has resulted in numerous drastic changes already being made, with some of these potentially lasting even after it is no longer a threat.

# Chapter 16: Conclusion

## Technology and the Pandemic

Since the start of the COVID-19 pandemic, the virus has wreaked havoc on a multitude of communities, cities, and countries. At the time that this book has been written, approximately 20 million individuals were diagnosed with COVID-19 worldwide and approximately 730,000 deaths have followed since the beginning of its spread (Worldometer, 2020). Beyond these consequences, the pandemic will have lasting impacts for years to come on different generations whose effects could last a lifetime. In Canada alone, approximately 30% of Generation Z individuals have lost their jobs during the pandemic and 10% of Canadian students said that they were unable to complete their degree as planned (Saba, 2020). However, amidst the chaos in diagnosing, treating, and containing the virus while seeking ways to adjust to the "new normal", technology has had an essential role in all aspects of the COVID-19 pandemic.

From the beginning, technology has been deeply intertwined with the daily lives of modern society. However, this interconnectedness, which aimed to increase convenience and productivity within and between different parties, has become both humanities' biggest strength and biggest weakness. The novel coronavirus spread like wildfire amidst the new age of increased air travel and the plague of harmful misinformation spread similarly through the continuously growing platform of social media users. Technology, in some cases, expedited the detrimental outcomes of the virus more rapidly in comparison to previous pandemics in which there was a lesser dependence on technology. Alternatively, the increased presence of technology in everyday life has also helped aid the fight against the novel virus. Certain technology is used to protect patients, healthcare providers, and other citizens while other types are used to seek treatments, provide therapy, or simply keep each other connected. The evaluations and assessments surrounding the role of technology in the COVID-19 pandemic demonstrates technology's prevalence in all aspects of life and its importance in battling a viral outbreak.

## Pre-existing and Newly Developed Technology

Existing modern technology has equipped humanity with several tools to fight the current pandemic more efficiently and more rapidly than those in the past. Refined models of personal protective equipment (PPE) such as gloves, masks, and gowns have been continuously improving throughout the years to better protect both the patient and the healthcare provider. Similar technology that aids in the efforts against the pandemic include diagnostic technologies and intensive care unit (ICU) support such as ventilators, however, the COVID-19 pandemic has greatly exposed the shortcomings of existing technology.

A multitude of shortages and breakages in the supply chain for essential technology call for more innovative methods to procure efficient and reliable alternatives for both the coveted tools and its production. Equipment vital to patient and medical staff safety, diagnostic devices, and research into vaccines as cures or treatments have all undergone significant innovation since the start of the pandemic in order to slow or stop its spread. For example, as different industries join forces to aid in efforts against the virus, manufacturers of automotives have begun to direct their production towards supplying ventilators for local hospitals (Layson, 2020). Furthermore, individuals with 3D printers are also helping through crowdsourcing programs to print masks and other equipment in short supply (The Top 5 Practical Digital Health Technologies in the Fight Against COVID-19, 2020). To battle shortages in ICU resources and availability, hospitals and institutions have developed alternatives for the traditional ventilator that allow medical staff to direct these supplies to those who are most in need (Kunzmann, 2020; MIT Emergency Ventilator, n.d). Testing has also improved as antibody tests and immunohistochemistry have allowed for more efficient and more accurate testing for the COVID-19 virus in individuals. Finally, in hopes of finding a cure, innovation from previously existing technology has allowed the determination of the novel virus' genome and thereby accelerates the search for an effective vaccine (Science Daily, 2020).

The existing technology in the medical field has since come a long way since the H1N1 outbreak in 2009 and even more since the 1918 Influenza pandemic. The fight against the COVID-19 pandemic would have unfolded very differently in the absence of technology and would have been futile without modern day advancements and innovation. Through the comparison of pre-existing technology and the development of newer technology, the necessity of innovation is illuminated and demonstrates the need for the continuation of technological research and development.

**Physical and Psychological Well-being**

As previously explored in this book, social distancing and isolation regulations had been enacted by governments worldwide to prevent the spread of the pandemic. The consequences that followed included adverse psychological and non-virus related physical health effects in many individuals, with particular focus on the elderly populations. However, with the advancement of technology, the era of video and audio calls, instant messaging, and social media have provided an unprecedented luxury of staying connected during these times of isolation. These platforms, supported by the prevalence of technology in everyday life, have been crucial to preventing the deterioration of mental health and have even been able to boost physical health for some. Content creators of home workouts, do-it-yourself videos, and other boredom fighters have galvanized generations to create and maintain a healthy lifestyle, even when going outside is not possible, and to stray away from a sedentary way of living.

Especially in light of social distancing and isolation, the absence of technology throughout the COVID-19 pandemic would have resulted in lasting detrimental effects on mental health and physical well-being. Technology has not only allowed individuals to remain connected, but also to adapt amidst change. Gym and community centre closures have been converted to online workout sessions and meeting with friends have become group calls; even in a pandemic, technology provides individuals with ways to support their mental and physical health.

However, while these resources are available to many, it is important to note that these resources must be seized in order to be efficacious. Only when the effort is taken to reach out to loved ones, communicate with friends, and to search for productivity videos can technology be used as an asset in the midst of maintaining one's physical and psychological well-being.

**Demands and Dependence on Technology**

Since the invention of the wheel, humanity has co-evolved with the advancement and continuous creation of technology. This creates a simultaneous demand for both increased quantity and quality of technology as well as a dependence on the technology created. However, as technology has created many conveniences in everyday life, much of society would crumble in its absence.

Society's vulnerability in the absence of technology was strikingly revealed through the events of the pandemic primarily through the shortages of PPE, testing kits, and ventilators. These essential tools are used to ensure the safety of both healthcare workers and patients, but a lack of these vital resources place both parties in danger. Hospitals situated in pandemic epicentres were forced to decide which COVID-19 patients would be fortunate enough to gain admission to the limited ICU space and which patients would be left to fate (Beall, 2020). Rural hospitals have also struggled during the pandemic with increased mortality rates since, historically, high mortality has been associated with a shortage of available equipment and resources (Kirakli et al., 2011). Throughout the course of the pandemic, the virus has made it clear that the world was unable to support its own dependence, and subsequent demand, on life-saving technology.

This finding illustrates the consequent demand on continuous and rapid innovation of technology. While the COVID-19 pandemic wreaked havoc across the globe, future catastrophes could include war, natural disasters, and the impending issue of global warming which will all require more quantities of improved and developed technology to solve. However, not only is there a demand to develop new technology, but also a constant need to continue improving the existing devices humanity depends on and its production methods of all sorts of technology to reduce the morbidity and mortality of pandemics or other disasters in the future.

## Socio-political Consequences

The influence of technology on everyday life has also had social, political, and economic impacts amidst the pandemic. Different technology platforms that enable teleconferencing have sparked the increase of individuals that work remotely from home. As well, these same platforms have also offered alternatives to students in the form of online learning that allows them to continue their education amidst school and campus closures. Subsequently, these companies pertaining to the fields of telecommunication and other necessary technologies during the pandemic, such as those that produce PPE or ventilators, are thriving in the current economic atmosphere.

Unfortunately, small businesses of all industries are struggling due to social distancing regulations and forced closures. To combat this, governments have been imposing policies that prevent monopolistic behaviour in the tech industry (Lewis, 2020). Other laws that have been influenced by technology include those that have been put in place to prevent the spread of misinformation. False claims and stories have the potential to spread like wildfire in the new age of social media and accessible internet and have galvanized governments to take action to contain the deceitful information from circulating in the public (Thompson, 2020). Ultimately, technology has reshaped the way that governments and businesses are managed and also influence the nature of policies and regulations that are put in place during a global crisis.

## Impacts of Telecommunication, the Internet, and Media

The COVID-19 pandemic differs from previous global outbreaks due to the increasing social media presence in everyday life. With the number of active users continuously growing on a variety of media platforms, the spread of rumours, false facts, and misinformation proceeds at an unprecedented rate. This circulation of untrue information proves counterproductive in efforts combating the novel virus as many individuals are exposed to more false claims than credible ones. Consequently, large populations refuse to abide by public health recommendations and further facilitates the spread of the virus.

Alternatively, the rise of technology in the form of social media and telecommunication in recent years also has the potential to connect individuals during these times of isolation. As previously mentioned, connectivity between individuals is crucial in preserving mental health and also allows for individuals to continue with previous commitments before the pandemic such as school or work. While technology may sometimes prove a weakness amidst a global crisis, it can also serve as one of humanities' greatest assets if used correctly.

**Long Term Consequences**

As the course of the pandemic continues, and even begins to settle, many individuals will be experiencing what has been referred to as a "pandemic hangover" for many years to come. While these changes will be experienced on individual, national, and global scales, humanity will be forced to adopt a new normal during the recovery period of the pandemic. Examples of changes on global and national scales may include efficiency focused global supply chains and strategies to prepare for future pandemics and emergencies. For individuals, impacts may include more remote work or schooling and a changed perspective on the ill. However, amidst the inevitable adjustment period, technology and its innovation will undoubtedly help society adapt and recover from the effects of the COVID-19 pandemic and prepare for the future.

**An Evaluation of Technology and its Role in the COVID-19 Pandemic**

Throughout this book, a variety of viewpoints and perspectives have been discussed regarding technology and its role in different aspects of the pandemic. From medical technologies on the frontlines of care to telecommunication, the internet, and social media to help those staying home, technology has been a crucial part in persisting since the pandemic's beginning. The effects of technology on humanity throughout this global crisis have made it clear that the outcome of the

COVID-19 pandemic would have been drastically different in its absence. Especially when comparing this outbreak with previous pandemics in history, the use of technology has had both its benefits and its drawbacks in the fight against the virus.

However, while the use of technology has had negative effects in terms of the spread of misinformation, the absence of technology, with regards to PPE and other essential equipment, has proved to have much dire consequences. Technology was a vital tool in battling the COVID-19 pandemic which ultimately demonstrates the need to invest in more technology, both in terms of existing and new technology, and create better production methods that are efficient, reliable, and sustainable in order to better equip humanity to tackle global crises in the future.

# References

Abbas, W. (2020, June 2). Coronavirus Impact: Air Travel May Return to Normal by 2021 Summer, says Emirates President. Retrieved July 29, 2020, from https://www.aviationpros.com/airlines/news/21140520/coronavirus-impact-air-travel-may-return-to-normal-by-2021-summer-says-emirates-president

Aguiar, Eric R. G. R., et al. "The COVID-19 Diagnostic Technology Landscape: Efficient Data Sharing Drives Diagnostic Development." Frontiers in Public Health, vol. 8, 2020, doi:10.3389/fpubh.2020.00309.

Ahmad, T. (2020). Corona Virus (COVID-19) Pandemic and Work from Home: Challenges of Cybercrimes and Cybersecurity. SSRN Electronic Journal. https://doi.org/10.2139/ssrn.3568830

Aiello, R. (2020, July 31). COVID-19 exposure notification app now available. Retrieved August 7, 2020, from https://www.ctvnews.ca/health/coronavirus/covid-19-exposure-notification-app-now-available-1.5046868

Air Canada. (2020, July 8). Air Canada to Explore Rapid COVID-19 Testing with Spartan Bioscience. Retrieved July 29, 2020, from https://aircanada.mediaroom.com/2020-07-08-Air-Canada-to-Explore-Rapid-COVID-19-Testing-with-Spartan-Bioscience

Alang, N. (2020, July 4). While most businesses suffer, the pandemic has only increased the frightening dominance of Facebook, Apple, Google and other giants | The Star. Retrieved August 10, 2020, from https://www.thestar.com/business/opinion/2020/07/04/big-tech-as-its-own-virus-covid-19-has-only-increased-the-reach-of-facebook-apple-google-and-other-giants.html

Amlc, Y. (2020, May 19). COVID 19's Impact on the Technology and Software Sector | BMO Capital Markets. Retrieved August 10, 2020, from https://capitalmarkets.bmo.com/en/news-insights/covid-19-insights/technology-business-services/covid-19s-impact-technology-and-software-sector/

Andersen, A. L., Hansen, E. T., Johannesen, N., & Sheridan, A. (2020). Pandemic, Shutdown and Consumer Spending: Lessons from Scandinavian Policy Responses to COVID-19. Retrieved from http://arxiv.org/abs/2005.04630

App Annie. (2020). Weekly Time Spent in Apps Grows 20% Year Over Year as People Hunker Down at Home. Retrieved July 31, 2020, from https://www.appannie.com/en/insights/market-data/weekly-time-spent-in-apps-grows-20-year-over-year-as-people-hunker-down-at-home/

Aydin, Suleyman. "A Short History, Principles, and Types of ELISA, and Our Laboratory Experience with Peptide/Protein Analyses Using ELISA." Peptides, vol. 72, 2015, pp. 4–15., doi:10.1016/j.peptides.2015.04.012.

Bahl, P., Doolan, C., Silva, C. D., Chughtai, A. A., Bourouiba, L., & Macintyre, C. R. (2020).

Airborne or Droplet Precautions for Health Workers Treating Coronavirus Disease 2019? The Journal of Infectious Diseases. doi:10.1093/infdis/jiaa189

Baig, A., Hall, B., Jenkins, P., Lamarre, E., & McCarthy, B. (2020, May 14). Digital adoption through COVID-19 and beyond | McKinsey. Retrieved August 8, 2020, from https://www.mckinsey.com/business-functions/mckinsey-digital/our-insights/the-covid-19-recovery-will-be-digital-a-plan-for-the-first-90-days

Bakx, K. (2020, April 29). How new technologies could protect industrial workers from COVID-19 | CBC News. Retrieved August 10, 2020, from https://www.cbc.ca/news/business/industrial-workplace-covid-19-1.5548516

Ball, P., & Maxmen, A. (2020, May 27). The epic battle against coronavirus misinformation and conspiracy theories. Retrieved August 8, 2020, from https://www.nature.com/articles/d41586-020-01452-z

Barnes, J. E. (2020, July 16). Russian Hackers Trying to Steal Coronavirus Vaccine Research - The New York Times. Retrieved August 8, 2020, from https://www.nytimes.com/2020/07/16/us/politics/vaccine-hacking-russia.html

Bateman, C. (2007). Paying the price for AIDS denialism. South African Medical Journal, 97(10), 912–914.

Baumeister, R. F., & Leary, M. R. (1995). The Need to Belong: Desire for Interpersonal Attachments as a Fundamental Human Motivation. Psychological Bulletin, 117(3), 497–529. https://doi.org/10.1037/0033-2909.117.3.497

Beall, A. (2020, April 28). The heart wrenching choice of who lives and dies. Retrieved August 7, 2020, from https://www.bbc.com/future/article/20200428-coronavirus-how-doctors-choose-who-lives-and-dies

Beglin, C. K. (2020, May 17). Why Was Mask Wearing Popular In Asia Even Before Covid-19? Retrieved July 30, 2020, from https://www.psychologytoday.com/us/blog/culture-shocked/202005/why-was-mask-wearing-popular-in-asia-even-covid-19

Belliveau, J. (2017, January 25). Wrist Arthritis: Symptoms, Treatment, and More. Retrieved July 31, 2020, from https://www.healthline.com/health/arthritis-wrist

Benstead, S. (2019, November 28). A four-day week: Is it really worth it? Retrieved July 27, 2020, from https://www.breathehr.com/blog/the-four-day-work-week-productive-or-pointless

Benveniste, A. (2020, April 1). The $94 billion fitness industry is reinventing itself as Covid-19 spreads - CNN. Retrieved July 31, 2020, from https://www.cnn.com/2020/04/01/business/fitness-studios-coronavirus/index.html

Bhatraju, P. K., Ghassemieh, B. J., Nichols, M., Kim, R., Jerome, K. R., Nalla, A. K., Greninger, A. L., Pipavath, S., Wurfel, M. M., Evans, L., Kritek, P. A., West, T. E., Luks, A., Gerbino, A., Dale, C. R., Goldman, J. D., O'Mahony, S., & Mikacenic, C. (2020). COVID-19 in critically ill patients in the Seattle region — Case series. New England Journal of Medicine, 382(21), 2012–2022. https://doi.org/10.1056/NEJMoa2004500

Bialek, S., Boundy, E., Bowen, V., Chow, N., Cohn, A., Dowling, N., Ellington, S., Gierke, R., Hall, A., MacNeil, J., Patel, P., Peacock, G., Pilishvili, T., Razzaghi, H., Reed, N., Ritchey, M., & Sauber-Schatz, E. (2020). Severe Outcomes Among Patients with Coronavirus Disease 2019 (COVID-19) — United States, February 12–March 16, 2020. MMWR. Morbidity and Mortality Weekly Report, 69(12), 343–346. https://doi.org/10.15585/mmwr.mm6912e2

Binder, D. (2020, April 20). COVID-19 Supply Chain Resources & Strategies - Inbound Logistics. Retrieved August 10, 2020, from https://www.inboundlogistics.com/cms/article/covid-19-supply-chain-resources-and-strategies/

Bleyleben, M. (2020, April 21). What every kids app developer should be doing during Covid-19 - SuperAwesome. Retrieved July 31, 2020, from https://www.superawesome.com/blog/reacting-to-the-explosion-in-kids-what-every-app-developer-should-be-doing-during-covid-19/

Bonner, A. (2019). What is Deep Learning and How Does it Work? Retrieved July 23, 2020, from https://towardsdatascience.com/what-is-deep-learning-and-how-does-it-work-f7d02aa9d477

Boseley, S. (2020, May 22). Hydroxychloroquine: Trump's Covid-19 "cure" increases deaths, global study finds | Science | The Guardian. Retrieved August 8, 2020, from https://www.theguardian.com/science/2020/may/22/hydroxychloroquine-trumps-covid-19-cure-increases-deaths-global-study-finds

Breen, G. M., & Matusitz, J. (2010). An evolutionary examination of telemedicine: A health and computer-mediated communication perspective. In Social Work in Public Health (Vol. 25, Issue 1, pp. 59–71). NIH Public Access. https://doi.org/10.1080/19371910902911206

Brown, D. (2020, April 26). Toronto synagogue's Zoom prayer service hijacked by trolls screaming derogatory slurs | CBC News. Retrieved August 9, 2020, from https://www.cbc.ca/news/canada/toronto/zoom-church-service-hijacked-toronto-synagogue-hate-crime-1.5545776

Bucher, J. T., & Cooper, J. S. (2018). Bag Mask Ventilation (Bag Valve Mask, BVM). In StatPearls. StatPearls Publishing. http://www.ncbi.nlm.nih.gov/pubmed/28722953

Byford, S. (2020, April 30). Coronavirus causes worst smartphone market contraction in history - The Verge. Retrieved August 10, 2020, from https://www.theverge.com/2020/4/30/21243685/coronavirus-smartphone-market-impact-q1-2020-covid-19

Callaway, E. (2020). The race for coronavirus vaccines: A graphical guide. Nature, 580(7805), 576-577. doi:10.1038/d41586-020-01221-y

Cassidy, J. T., & Baker, J. F. (2016). Orthopaedic Patient Information on the World Wide Web. The Journal of Bone and Joint Surgery, 98(4), 325–338. https://doi.org/10.2106/jbjs.n.01189

Centre for Disease Control. (2020, May 8). Benefits of Physical Activity | Physical Activity | CDC. Retrieved July 31, 2020, from https://www.cdc.gov/physicalactivity/basics/pa-health/index.htm

Chan, J. F., Yip, C. C., & Tai, K. K. (2020, April 23). Improved Molecular Diagnosis of COVID-19 by the Novel, Highly Sensitive and Specific COVID-19-RdRp/Hel Real-Time Reverse Transcription-PCR Assay Validated In Vitro and with Clinical Specimens. Retrieved August 11, 2020, from https://pubmed.ncbi.nlm.nih.gov/32132196/

Chandler, G. N., Keller, C., & Lyon, D. W. (2000). Unraveling the Determinants and Consequences of an Innovation-Supportive Organizational Culture. Entrepreneurship Theory and Practice, 25(1), 59–76. https://doi.org/10.1177/104225870002500106

Chen, W. (2020). Promise and challenges in the development of COVID-19 vaccines. Human Vaccines & Immunotherapeutics, 1-5. doi:10.1080/21645515.2020.1787067

Christophersen, O. A., & Haug, A. (2006). Why is the world so poorly prepared for a pandemic of hypervirulent avian influenza? Microbial Ecology in Health & Disease, 18(3), 4th ser., 113-132. doi:10.3402/mehd.v18i3-4.7678

Cinelli, M., Quattrociocchi, W., Galeazzi, A., Valensise, C. M., Brugnoli, E., Schmidt, A. L., … Scala, A. (2020). The COVID-19 Social Media Infodemic.

Clark, D. A. (2020, March 3). Media, Fear, and the Coronavirus Outbreak. https://www.psychologytoday.com/ca/blog/the-runaway-mind/202003/media-fear-and-the-coronavirus-outbreak.

Cohen, J. (2020). The $1 billion bet: Pharma giant and U.S. government team up in all-out coronavirus vaccine push. Science. https://doi.org/10.1126/science.abc0056

Crabtree, J. (2020, March 25). Coronavirus crisis will send globalization into reverse. Retrieved July 28, 2020, from https://asia.nikkei.com/Opinion/Coronavirus-crisis-will-send-globalization-into-reverse?utm_source=Eurasia+Group+Signal

Cuan-Baltazar, J. Y., Muñoz-Perez, M. J., Robledo-Vega, C., Pérez-Zepeda, M. F., & Soto-Vega, E. (2020). Misinformation of COVID-19 on the Internet: Infodemiology Study (Preprint). Journal of Medical Internet Research, 6(2). https://doi.org/10.2196/preprints.18444

CYCADSG. (2019). Technology Influence on Recreation and Leisure Time – cycadsg.org. Retrieved July 31, 2020, from https://cycadsg.org/technology-influence-on-recreation-and-leisure-time/

Daley, S. (2019). What is Artificial Intelligence? How Does AI Work? | Built In. Retrieved July 23, 2020, from https://builtin.com/artificial-intelligence

Davidson, H. (2020). Around 20% of global population under coronavirus lockdown | World news | The Guardian. Retrieved July 31, 2020, from https://www.theguardian.com/world/2020/mar/24/nearly-20-of-global-population-under-coronavirus-lockdown

DeLaet, R., & Dubeau, E. (2020). Diagnosing COVID-19 using artificial intelligence | Lawson Health Research Institute. Retrieved July 23, 2020, from https://www.lawsonresearch.ca/news/diagnosing-covid-19-using-artificial-intelligence

Deloitte. (2020a). COVID-19: A black swan event for the semiconductor industry. Retrieved from https://www2.deloitte.com/global/en/pages/about-deloitte/articles/a-black-swan-event-for-the-semiconductor-industry-covid-19.html

Deloitte. (2020b). Understanding the sector impact of Technology sector.

Depoux, A., Martin, S., Karafillakis, E., Preet, R., Wilder-Smith, A., & Larson, H. (2020). The pandemic of social media panic travels faster than the COVID-19 outbreak. Journal of Travel Medicine, 27(3). https://doi.org/10.1093/jtm/taaa031

Desai, S., Mooney, H., & Oehrli, J. A. (2020, June 24). What is "Fake News"? - "Fake News," Lies and Propaganda: How to Sort Fact from Fiction - Research Guides at University of Michigan Library. Retrieved August 8, 2020, from https://guides.lib.umich.edu/fakenews

Dietz, W., & Santos-Burgoa, C. (2020). Obesity and its Implications for COVID-19 Mortality. Obesity, 28(6), 1005–1005. https://doi.org/10.1002/oby.22818

Dingel, J. I., & Neiman, B. (2020). How Many Jobs Can be Done at Home? https://github.com/jdingel/DingelNeiman-workathome.

DISCERN. DISCERN Background. http://www.discern.org.uk/background_to_discern.php.

Donnelly, N., & Proctor-Thomson, S. B. (2015). Disrupted work: home-based teleworking (HbTW) in the aftermath of a natural disaster. New Technology, Work and Employment, 30(1), 47–61. https://doi.org/10.1111/ntwe.12040

Draper, S. J., & Heeney, J. L. (2009). Viruses as vaccine vectors for infectious diseases and cancer. Nature Reviews Microbiology, 8(1), 62-73. doi:10.1038/nrmicro2240

Eckert, A. (n.d.). A surprising player in the race for a SARS-CoV-2 vaccine. Retrieved July 26, 2020, from https://www.nature.com/articles/d42473-020-00032-z

Edwards, E. (n.d.). What is Melt-Blown Extrusion and How is it Used for Making Masks? Retrieved July 16, 2020, from https://www.thomasnet.com/articles/machinery-tools-supplies/what-is-melt-blown-extrusion/

Emarysys. (2020). US retailers see online growth YoY in April similar to recent holiday season - Covid-19 World Commerce Impact. Retrieved August 10, 2020, from https://ccinsight.org/observations/us-retailers-see-online-growth-yoy-in-april-similar-to-recent-holiday-season/

Erdelyi, K. M. (2019, October 21). The Psychological Impact of Information Warfare & Fake News. Retrieved August 09, 2020, from https://www.psycom.net/iwar.1.html

Erwin, C., Aultman, J., Harter, T., Illes, J., & Kogan, R. C. J. (2020). Rural and Remote Communities: Unique Ethical Issues in the COVID-19 Pandemic. In American Journal of Bioethics. Routledge. https://doi.org/10.1080/15265161.2020.1764139

Expertsystem. (2017). What is Machine Learning? A definition - Expert System. Retrieved July 23, 2020, from https://expertsystem.com/machine-learning-definition/

Fadare, O. O., & Okoffo, E. D. (2020). Covid-19 face masks: A potential source of microplastic fibers in the environment. Science of The Total Environment, 737. doi:10.1016/j.scitotenv.2020.140279

Fetterly, R. (2020, May). Planning for After the COVID-19 Pandemic. Retrieved July 27, 2020, from https://www.cgai.ca/planning_for_after_the_covid_19_pandemic

Fildes, N., & Espinoza, J. (2020, April 8). Tracking coronavirus: big data and the challenge to privacy . Financial Times. https://www.ft.com/content/7cfad020-78c4-11ea-9840-1b8019d9a987

Fischer, S. (2020, March 21). Kids' daily screen time surges during coronavirus - Axios. Retrieved July 31, 2020, from https://www.axios.com/kids-screen-time-coronavirus-562073f6-0638-47f2-8ea3-4f8781d6b31b.html

Forster, V. (2020, June 1). Are Your Eyes Hurting During The Coronavirus Pandemic? You May Have "Computer Vision Syndrome." Retrieved July 31, 2020, from https://www.forbes.com/sites/victoriaforster/2020/06/01/are-your-eyes-hurting-during-the--coronavirus-pandemic-you-may-have-computer-vision-syndrome/#424a6b246cf9

Fraumeni, P. (2020). How AI is enabling COVID-19 research. Retrieved July 23, 2020, from https://www.utoronto.ca/news/how-ai-enabling-covid-19-research

Gao, Yan, et al. "Structure of the RNA-Dependent RNA Polymerase from COVID-19 Virus." Science, American Association for the Advancement of Science, 15 May 2020, science.sciencemag.org/content/368/6492/779.

Garrity, K. (2020). Northeastern MOBS Lab researches Coronavirus - The Huntington News. Retrieved July 23, 2020, from https://huntnewsnu.com/61923/campus/northeastern-mobs-lab-researches-coronavirus/

Gopinath, G. (2020, April 14). The Great Lockdown: Worst Economic Downturn Since the Great Depression. Retrieved July 28, 2020, from https://blogs.imf.org/2020/04/14/the-great-lockdown-worst-economic-downturn-since-the-great-depression/

Grady, D. (2020, April 20). The Pandemic's Hidden Victims: Sick or Dying, but Not From the Coronavirus - The New York Times. New York Times. https://www.nytimes.com/2020/04/20/health/treatment-delays-coronavirus.html

Guyot, K., & Sawhill, I. V. (2020, April 6). Telecommuting will likely continue long after the pandemic. Brookings. https://www.brookings.edu/blog/up-front/2020/04/06/telecommuting-will-likely-continue-long-after-the-pandemic/

Hackett, R. (2016, August 12). Area 1 Report Reveals How Hackers Hide Their Tracks | Fortune. Retrieved August 9, 2020, from https://fortune.com/2016/08/12/how-hackers-hide-tracks-cyberattacks/

Halupa, C. (2016). RISKS: THE IMPACT OF ONLINE LEARNING AND TECHNOLOGY ON STUDENT PHYSICAL, MENTAL, EMOTIONAL, AND SOCIAL HEALTH. ICERI2016 Proceedings, 1, 6305–6314. https://doi.org/10.21125/iceri.2016.0044

Harrington, S. C., Stack, J., & O'dwyer, V. (2019). Risk factors associated with myopia in schoolchildren in Ireland. British Journal of Ophthalmology, 103(12), 1803–1809. https://doi.org/10.1136/bjophthalmol-2018-313325

Haselton, T. (2020, April 27). Apple's iPhone 12 production delayed, report says. Retrieved August 10, 2020, from

118

https://www.cnbc.com/2020/04/27/apples-iphone-12-production-delayed-report-says.html

Healthwise Staff. (2019, June 26). Proper Sitting Posture for Typing. Retrieved July 31, 2020, from https://myhealth.alberta.ca/Health/Pages/conditions.aspx?hwid=hw200906

Heaton, P. M. (2020). The Covid-19 Vaccine-Development Multiverse. The New England Journal of Medicine, NEJMe2025111. https://doi.org/10.1056/NEJMe2025111

Hecht, M. (2017, March 13). Repetitive Strain Injury (RSI): Causes, Prevention, and More. Retrieved July 31, 2020, from https://www.healthline.com/health/repetitive-strain-injury

Hendry, M. (2020, June 12). COVID-19: How technology can save your workforce. Retrieved August 10, 2020, from https://www.hcamag.com/ca/specialization/hr-technology/covid-19-how-technology-can-save-your-workforce/225147

Henneberry, B. (n.d.). How to Make N95 Masks. Retrieved July 15, 2020, from https://www.thomasnet.com/articles/plant-facility-equipment/how-to-make-n95-masks/

Hensley, L. (2020, April 12). What is a ventilator, and why do some coronavirus patients need one? - National | Globalnews.ca. https://globalnews.ca/news/6787773/what-are-ventilators-coronavirus/

Hinton, G. E., Osindero, S., & Teh, Y. W. (2006). A fast learning algorithm for deep belief nets. Neural Computation, 18(7), 1527–1554. https://doi.org/10.1162/neco.2006.18.7.1527

Hormann, C., Baum, M., Putensen, C., Mutz, N. J., & Benzer, H. (1994). Biphasic positive airway pressure (BIPAP) - a new mode of ventilatory support. European Journal of Anaesthesiology, 11(1), 37–42. http://europepmc.org/article/med/8143712

Huang, H.-M., Chang, D. S.-T., & Wu, P.-C. (2015). The Association between Near Work Activities and Myopia in Children—A Systematic Review and Meta-Analysis. PLOS ONE, 10(10), e0140419. https://doi.org/10.1371/journal.pone.0140419

Ikuno, Y. (2017). OVERVIEW OF THE COMPLICATIONS OF HIGH MYOPIA. Retina, 37(12), 2347–2351. https://doi.org/10.1097/IAE.0000000000001489

Ishack, S., & Lipner, S. R. (2020). Applications of 3D Printing Technology to Address COVID-19–Related Supply Shortages. The American Journal of Medicine, 133(7), 771-773. doi:10.1016/j.amjmed.2020.04.002

Italie, L. (2020, May 26). Worry, haste, retail therapy: What have we bought and why? | CTV News. Retrieved August 10, 2020, from https://www.ctvnews.ca/health/coronavirus/worry-haste-retail-therapy-what-have-we-bought-and-why-1.4954991

Jackson, C. D. (2019, April 11). What is intrinsic positive end-expiratory pressure (PEEP), or auto-PEEP, in mechanical ventilation? https://www.medscape.com/answers/304068-104803/what-is-intrinsic-positive-end-expiratory-pressure-peep-or-auto-peep-in-mechanical-ventilation

Jasinski, G. (2020, May 19). Supplyframe Electronics Sourcing Report Highlights Innovation Imperative Amid COVID-19 | Business Wire. Retrieved August 10, 2020, from https://www.businesswire.com/news/home/20200519005327/en/Supplyframe-Electronics-Sourcing-Report-Highlights-Innovation-Imperative

John, T., & Wedeman, B. (2020, March 8). Italy prohibits travel and cancels all public events in its northern region to contain coronavirus. https://edition.cnn.com/2020/03/08/europe/italy-coronavirus-lockdown-europe-intl/index.html.

Johnson, J. (2020, February 25). Negative effects of technology: Psychological, social, and health. Retrieved July 31, 2020, from https://www.medicalnewstoday.com/articles/negative-effects-of-technology

Jones, D. (2020, May 29). These destinations will basically pay you to come visit during the pandemic. Retrieved July 29, 2020, from https://www.washingtonpost.com/travel/2020/05/29/these-destinations-will-basically-pay-you-come-visit-during-pandemic/

Jones, R. P. (2020, June 12). Airline temperature checks to become mandatory for travellers in Canada | CBC News. Retrieved July 29, 2020, from https://www.cbc.ca/news/politics/airline-passenger-temperature-checks-1.5609564

Jumper, J., Tunyasuvanakool, K., Kohli, P., & Hassabis, D. (2020). Computational predictions of protein structures associated with COVID-19 | DeepMind. Retrieved July 23, 2020, from https://deepmind.com/research/open-source/computational-predictions-of-protein-structures-associated-with-COVID-19

Kanda, W., & Kivimaa, P. (2020). What opportunities could the COVID-19 outbreak offer for sustainability transitions research on electricity and mobility? Energy Research & Social Science, 68, 101666. doi:10.1016/j.erss.2020.101666

Kelly, M. (2020, July 28). Tech's four biggest companies are going on trial - The Verge. Retrieved August 8, 2020, from https://www.theverge.com/2020/7/28/21344920/big-tech-ceo-antitrust-hearing-apple-facebook-amazon-google-facebook

Kelly, T. (2020, June 23). How hackers are using COVID-19 to find new phishing victims | 2020-06-23 | Security Magazine. Retrieved August 8, 2020, from https://www.securitymagazine.com/articles/92666-how-hackers-are-using-covid-19-to-find-new-phishing-victims

Kim, A. (2020, July 04). MIT-designed robot can disinfect a warehouse floor in 30 minutes -- and could one day be employed in grocery stores and schools. Retrieved August 10, 2020, from https://www.cnn.com/2020/07/04/tech/mit-csail-coronavirus-robot-scn-trnd/index.html

Kirakli, C., Tatar, D., Cimen, P., Edipoglu, O., Coskun, M., Celikten, E., & Ozsoz, A. (2011). Survival from severe pandemic H1N1 in urban and rural turkey: A case series. Respiratory Care, 56(6), 790–795. https://doi.org/10.4187/respcare.00988

Koeze, E., & Popper, N. (2020, April 7). The Virus Changed the Way We Internet - The New York Times. Retrieved July 31, 2020, from https://www.nytimes.com/interactive/2020/04/07/technology/coronavirus-internet-use.html

Kohl, H. W., & Cook, H. D. (2013). Educating the Student Body. In Educating the Student Body. https://doi.org/10.17226/18314

Kose, M. A., Ohnsorge, F., Nagle, P., & Sugawara, N. (2020, June). COVID-19 and Debt Crises in Developing Economies - IMF F&D. Retrieved July 28, 2020, from https://www.imf.org/external/pubs/ft/fandd/2020/06/COVID19-and-debt-in-developing-economies-kose.htm

Kouzy, R., Jaoude, J. A., Kraitem, A., Alam, M. B. E., Karam, B., Adib, E., ... Baddour, K. (2020). Coronavirus Goes Viral: Quantifying the COVID-19 Misinformation Epidemic on Twitter. Cureus, 12(3). https://doi.org/10.7759/cureus.7255

Kunzmann, K. (2020, April 17). COVID-19 Ventilator Innovation from Mount Sinai | HCPLive. https://www.mdmag.com/medical-news/covid19-ventilator-innovation-from-mount-sinai

Kupferschmidt, K. (2020). 'Vaccine nationalism' threatens global plan to distribute COVID-19 shots fairly. Science. https://doi.org/10.1126/science.abe0601

La Trobe University. (2015, May). Computer-related injuries - Better Health Channel. Retrieved July 31, 2020, from https://www.betterhealth.vic.gov.au/health/healthyliving/computer-related-injuries

Langton, K. (2020, May 30). China lockdown: How long was China on lockdown? | Travel News | Travel | Express.co.uk. Retrieved August 10, 2020, from https://www.express.co.uk/travel/articles/1257717/china-lockdown-how-long-was-china-lockdown-timeframe-wuhan

Larson, H. J. (2018, October 16). The biggest pandemic risk? Viral misinformation. https://www.nature.com/articles/d41586-018-07034-4.

Layson, G. (2020, March 27). Canadian suppliers team up to help produce 10,000 ventilators for Ontario. https://canada.autonews.com/coronavirus/canadian-suppliers-team-help-produce-10000-ventilators-ontario

Levin, D. (2020). Using artificial intelligence to diagnose COVID-19. Retrieved July 23, 2020, from https://medicalxpress.com/news/2020-05-artificial-intelligence-covid-.html

Levy, A. (2020, March 19). Tech's big five lost $1 trillion in market value in past month. Retrieved August 10, 2020, from https://www.cnbc.com/2020/03/19/techs-big-five-lost-1-trillion-in-market-value-in-past-month.html

Lewis, A. (2020, April 28). Elizabeth Warren, Alexandria Ocasio-Cortez want mergers halted due to COVID-19 | PitchBook. Retrieved August 8, 2020, from https://pitchbook.com/news/articles/elizabeth-warren-alexandria-ocasio-cortez-want-mergers-halted-due-to-covid-19

Li, C. (2020). The COVID-19 pandemic has changed education forever. This is how. Retrieved August 10, 2020, from https://www.weforum.org/agenda/2020/04/coronavirus-education-global-covid19-online-digital-learning/

Liu, A. (2020, June 15). China's Sinovac plots pivotal COVID-19 vaccine trial in Brazil after positive phase 2. Retrieved July 27, 2020, from https://www.fiercepharma.com/vaccines/china-s-sinovac-says-covid-19-vaccine-shows-early-positive-results-phase-2

Lopez, C. T. (2020, June 10). Domestic N95 Mask Production Expected to Exceed 1 Billion in 2021. Retrieved July 13, 2020, from https://www.defense.gov/Explore/News/Article/Article/2215532/domestic-n95-mask-production-expected-to-exceed-1-billion-in-2021/

MacCharles, T. (2020, April 7). Canadian industries will make 30,000 ventilators for COVID-19 battle | The Star. https://www.thestar.com/politics/federal/2020/04/07/canadian-industries-will-make-30000-ventilators-for-covid-19-battle.html

Malone, K. G. (2020, March 25). Testing backlog linked to shortage of chemicals needed for COVID-19 test | CTV News. The Canadian Press. https://www.ctvnews.ca/health/coronavirus/testing-backlog-linked-to-shortage-of-chemicals-needed-for-covid-19-test-1.4867226

Manthous, C., & Tobin, M. (2017). American Thoracic Society PATIENT EDUCATION | INFORMATION SERIES Why are ventilators used? http://www.caregiver.org

Martinez, R. (2020, June 4). COVID-19 Drives Lasting Changes in Global Consumer Behavior and Businesses Operations | Deloitte | Blog. Retrieved August 10, 2020, from https://www2.deloitte.com/global/en/blog/responsible-business-blog/2020/covid-19-drives-lasting-changes-in-global-consumer-behavior-and-businesses-operations.html

Massey, D., & Cole, D. (2020, April 8). HHS to work with GM under Defense Production Act to produce 30,000 ventilators for national stockpile - CNNPolitics. https://www.cnn.com/2020/04/08/politics/general-motors-ventilators-defense-production-act-coronavirus/index.html

Mathur, N. (2020, July 01). How Technology Will Help Redefine The Workplace Post COVID-19. Retrieved August 10, 2020, from https://in.mashable.com/tech/15273/how-technology-will-help-redefine-the-workplace-post-covid-19

Mayo Clinic Staff. (2019, April 27). Office ergonomics: Your how-to guide - Mayo Clinic. Retrieved July 31, 2020, from https://www.mayoclinic.org/healthy-lifestyle/adult-health/in-depth/office-ergonomics/art-20046169

Mayo Clinic. (2020, May 28). COVID-19: How much protection do face masks offer? Retrieved July 14, 2020, from https://www.mayoclinic.org/diseases-conditions/coronavirus/in-depth/coronavirus-mask/art-20485449

McCabe, B. (2020, May 11). 3 Ways to Fix the Neck & Shoulder Pain You Feel While Working from Home - COVID-19, Health Topics, Pain Management, Physical Rehabilitation - Hackensack Meridian Health. Retrieved July 31, 2020, from https://www.hackensackmeridianhealth.org/HealthU/2020/05/11/3-ways-to-fix-the-neck-shoulder-pain-you-feel-while-working-from-home/

McCaffrey, M., Moline, R., & Palladino, J. T. (2020). COVID-19 and the technology Industry: PwC. Retrieved August 10, 2020, from https://www.pwc.com/us/en/library/covid-19/coronavirus-technology-impact.html

Mckay, J. (2020). AI-Based Location Positioning System for Private Contact Tracing. Retrieved July 23, 2020, from https://www.govtech.com/em/safety/-AI-Based-Location-Positioning-System-for-Private-Contact-Tracing.html

Mercurio, A. (2020, May 1). How COVID-19 impacts Indigenous communities - News and Events - Ryerson University. Ryerson Today. https://www.ryerson.ca/news-events/news/2020/05/how-covid-19-impacts-indigenous-communities/

Mian, A., & Khan, S. (2020). Coronavirus: the spread of misinformation. BMC Medicine, 18(1). https://doi.org/10.1186/s12916-020-01556-3

Milano, B. (2019, February 7). Government can't keep up with technology's growth – Harvard Gazette. Retrieved August 8, 2020, from https://news.harvard.edu/gazette/story/2019/02/government-cant-keep-up-with-technologys-growth/

Minaya, E. (2020, April 3). CFOs Plan To Permanently Shift Significant Numbers Of Employees To Work Remotely — Survey. https://www.forbes.com/sites/ezequielminaya/2020/04/03/cfos-plan-to-permanently-shift-significant-numbers-of-employees-to-work-remotely---survey/#1bbc78e575b2

MIT Emergency Ventilator. (n.d.). Retrieved July 12, 2020, from https://emergency-vent.mit.edu/

Mitter, S. (2020, June 08). BYJU'S founders on the rise of edtech amid COVID-19 and how the platform clocked Rs 350 Cr sales in a month. Retrieved

August 10, 2020, from https://yourstory.com/2020/06/byjus-founders-edtech-covid-19-startup

Morozov, Sergey. "Review of 'Antibody Detection and Dynamic Characteristics in Patients with COVID-19.'" 2020, doi:10.14322/publons.r7870744.

Murphy, S. (2011, February 22). Tech overload causing kids back, neck problems - Technology & science - Digital Home | NBC News. Retrieved July 31, 2020, from http://www.nbcnews.com/id/41718918/ns/technology_and_science-digital_home/t/tech-overload-causing-kids-back-neck-problems/#.XEEtMFwzbIU

National Health Service. (2018, November 19). Repetitive strain injury (RSI) - NHS. Retrieved July 31, 2020, from https://www.nhs.uk/conditions/repetitive-strain-injury-rsi/

National Sleep Foundation. (n.d.). How Technology Impacts Sleep Quality | Sleep.org. Retrieved July 31, 2020, from https://www.sleep.org/articles/ways-technology-affects-sleep/

Neergaard, L., & Fingerhut, H. (2020, May 27). AP-NORC poll: Half of Americans would get a COVID-19 vaccine. https://apnews.com/dacdc8bc428dd4df6511bfa259cfec44.

Neustaeter, B. (2020, July 07). Pandemic accelerating, global peak still to come: WHO chief. Retrieved August 3, 2020, from https://www.ctvnews.ca/health/coronavirus/pandemic-accelerating-global-peak-still-to-come-who-chief-1.5014246

Neustaeter, B. (2020, June 18). Canada expected to see spike in divorces as courts reopen, lawyers say. Retrieved July 29, 2020, from https://www.ctvnews.ca/health/coronavirus/canada-expected-to-see-spike-in-divorces-as-courts-reopen-lawyers-say-1.4989965

Novet, J. (2020, February 26). Zoom has added more users so far this year than in 2019: Bernstein. https://www.cnbc.com/2020/02/26/zoom-has-added-more-users-so-far-this-year-than-in-2019-bernstein.html

O'Hanlon, L. (n.d.). Are Virtual Elementary Schools Good for Kids? Retrieved August 10, 2020, from https://www.parents.com/kids/education/elementary-school/virtual-elementary-school-should-you-enroll-your-kids/

Ortiz-Ospina, E. (2019, September 18). The rise of social media. Retrieved August 5, 2020, from https://ourworldindata.org/rise-of-social-media

Page, S. (2020, July 07). The retired inventor of N95 masks is back at work, mostly for free, to fight covid-19. Retrieved August 3, 2020, from https://www.washingtonpost.com/lifestyle/2020/07/07/peter-tsai-n95-mask-covid/

Pan, Y., Long, L., Zhang, D., Yuan, T., Cui, S., Yang, P., . . . Ren, S. (2020). Potential False-Negative Nucleic Acid Testing Results for Severe Acute Respiratory Syndrome Coronavirus 2 from Thermal Inactivation of Samples with Low Viral Loads. *Clinical Chemistry, 66*(6), 794-801. doi:10.1093/clinchem/hvaa091

Pazzanese, C. (2020, May 8). Social media used to spread, create COVID-19 falsehoods. https://news.harvard.edu/gazette/story/2020/05/social-media-used-to-spread-create-covid-19-falsehoods/.

Perri, M., Dosani, N., & Hwang, S. W. (2020). COVID-19 and people experiencing homelessness: challenges and mitigation strategies. CMAJ : Canadian Medical Association Journal = Journal de l'Association Medicale Canadienne, 192(26), E716–E719. https://doi.org/10.1503/cmaj.200834

Perry, T. S. (2020). How Facebook Is Using AI to Fight COVID-19 Misinformation - IEEE Spectrum. Retrieved July 23, 2020, from https://spectrum.ieee.org/view-from-the-valley/artificial-intelligence/machine-learning/how-facebook-is-using-ai-to-fight-covid19-misinformation

Peters, D. J. (2020). Community Susceptibility and Resiliency to COVID-19 Across the Rural-Urban Continuum in the United States. The Journal of Rural Health, 36(3), 446–456. https://doi.org/10.1111/jrh.12477

Pinola, M. (2020, April 10). The remote worker's toolkit: The 15 tools you need to work remotely - The ultimate guide to remote work | Zapier. Retrieved August 10, 2020, from https://zapier.com/learn/remote-work/productivity-apps-remote-work/

Porpora, T. (2020, March 23). What is a ventilator? How much does one cost? - silive.com. https://www.silive.com/coronavirus/2020/03/what-is-a-ventilator-how-much-does-one-cost.html

Porter, J., & Ricker, T. (2020). Samsung and LG warn that the worst financial impact of COVID-19 is still to come - The Verge. Retrieved August 10, 2020, from https://www.theverge.com/2020/4/29/21240925/samsung-lg-earnings-q2-2020-coronavirus-earnings-down-tv-smartphones-memory-chips

Premier Orthopedics. (2015, January 22). Protecting your Wrists: 5 Exercise for Computer Users - Premier Orthopaedics. Retrieved July 31, 2020, from https://www.premierortho.com/protecting-your-wrist/protecting-wrists-5-exercise-computer-users/

Priest, D. (2020, August 07). Do air purifiers protect against the coronavirus? Retrieved August 10, 2020, from https://www.cnet.com/how-to/do-air-purifiers-protect-against-the-coronavirus/

Public Health Ontario. (2020). COVID-19 Contact Tracing Initiative | Public Health Ontario. Retrieved July 23, 2020, from https://www.publichealthontario.ca/en/diseases-and-conditions/infectious-diseases/respiratory-diseases/novel-coronavirus/contact-tracing-initiative

Ra, C. K., Cho, J., Stone, M. D., De La Cerda, J., Goldenson, N. I., Moroney, E., ... Leventhal, A. M. (2018). Association of digital media use with subsequent symptoms of attention-deficit/hyperactivity disorder among adolescents. JAMA - Journal of the American Medical Association, 320(3), 255–263. https://doi.org/10.1001/jama.2018.8931

Radcliffe, S. (2020, June 24). Where We're at with Vaccines and Treatments for COVID-19. Retrieved July 27, 2020, from https://www.healthline.com/health-news/heres-exactly-where-were-at-with-vaccines-and-treatments-for-covid-19

Ramdas, K., Darzi, A., & Jain, S. (2020). 'Test, re-test, re-test': using inaccurate tests to greatly increase the accuracy of COVID-19 testing. In Nature Medicine (Vol. 26, Issue 6, pp. 810–811). Nature Research. https://doi.org/10.1038/s41591-020-0891-7

Ranney, M. L., Griffeth, V., & Jha, A. K. (2020). Critical supply shortages - The need for ventilators and personal protective equipment during the Covid-19 pandemic. New England Journal of Medicine, 382(18), E41. https://doi.org/10.1056/NEJMp2006141

Reed, J. R. (2020). Tech companies pull back on hiring. Retrieved August 10, 2020, from https://www.cnbc.com/2020/04/23/tech-companies-pull-back-on-hiring.html

Reeves, R. V., & Rothwell, J. (2020, March 27). Class and COVID: How the less affluent face double risks. https://www.brookings.edu/blog/up-front/2020/03/27/class-and-covid-how-the-less-affluent-face-double-risks/

Ricketts, T. C. (2000). Health care in rural communities: The imbalance of health care resource distribution needs correction. In Western Journal of

Medicine (Vol. 173, Issue 5, pp. 294–295). BMJ Publishing Group. https://doi.org/10.1136/ewjm.173.5.294

Roberts, Y. (2020, April 29). Is The Online Fitness Boom Here To Stay? Retrieved July 31, 2020, from https://www.forbes.com/sites/yolarobert1/2020/04/29/is-the-online-fitness-boom-here-to-stay/#299173d47080

Robinson, J. (2019, May 12). Preventing Sleep Problems: Good Sleep Habits. Retrieved July 31, 2020, from https://www.webmd.com/sleep-disorders/preventing-sleep-problems

Rowan, N. J., & Laffey, J. G. (2020). Challenges and solutions for addressing critical shortage of supply chain for personal and protective equipment (PPE) arising from Coronavirus disease (COVID19) pandemic – Case study from the Republic of Ireland. Science of The Total Environment, 725. doi:10.1016/j.scitotenv.2020.138532

Saade, J. (2020, May 5). Technology and ethics in the coronavirus economy | TechCrunch. Retrieved August 8, 2020, from https://techcrunch.com/2020/05/05/technology-and-ethics-in-the-coronavirus-economy/

Saba, R. (2020, August 4). 'Life as we know it is over:' Why young Canadians will have a pandemic hangover for years to come. The Star. https://www.thestar.com/business/2020/08/04/life-as-we-know-it-is-over-why-young-canadians-will-have-a-pandemic-hangover-for-years-to-come.html

Sanger, D. E., & Perlroth, N. (2020, May 10). U.S. to Accuse China of Trying to Hack Vaccine Data, as Virus Redirects Cyberattacks - The New York Times. Retrieved August 8, 2020, from https://www.nytimes.com/2020/05/10/us/politics/coronavirus-china-cyber-hacking.html

Saunders, D. (2020, March 13). Hospital bed shortages amid COVID-19 expose the deadly cost of a lack of vision - The Globe and Mail. https://www.theglobeandmail.com/opinion/article-hospital-bed-shortages-amid-covid-19-expose-the-deadly-cost-of-a-lack/

Says, B., Says, H., Says, R., Says, T., Says, J., Says,, N. (2015, July 10). RNA vaccines: A novel technology to prevent and treat disease. Retrieved July 26, 2020, from http://sitn.hms.harvard.edu/flash/2015/rna-vaccines-a-novel-technology-to-prevent-and-treat-disease/

Schetzer, A. (2019, July 7). Governments are making fake news a crime – but it could stifle free speech. Retrieved August 8, 2020, from https://theconversation.com/governments-are-making-fake-news-a-crime-but-it-could-stifle-free-speech-117654

Schnurr, R. E., Alboiu, V., Chaudhary, M., Corbett, R. A., Quanz, M. E., Sankar, K., . . . Walker, T. R. (2018). Reducing marine pollution from single-use plastics (SUPs): A review [Abstract]. Marine Pollution Bulletin, 137, 157-171. doi:10.1016/j.marpolbul.2018.10.001

Science Daily. (2020, January 31). Whole genome of novel coronavirus, 2019-nCoV, sequenced. Retrieved August 4, 2020, from https://www.sciencedaily.com/releases/2020/01/200131114748.htm

Scroxton, A. (2020, June 17). Coronavirus: Cyber security spend to slow in 2020. Retrieved August 10, 2020, from https://www.computerweekly.com/news/252484783/Coronavirus-Cyber-security-spend-to-slow-in-2020

Seladi-Schulman, J. (2020, June 02). Can Face Masks Protect You from the 2019 Coronavirus? Retrieved July 15, 2020, from https://www.healthline.com/health/coronavirus-mask

Shankar, A., Hamer, M., McMunn, A., & Steptoe, A. (2013). Social isolation and loneliness: Relationships with cognitive function during 4 years of follow-up in the English longitudinal study of ageing. Psychosomatic Medicine, 75(2), 161–170. https://doi.org/10.1097/PSY.0b013e31827f09cd

Shearer, E. (2018, December 10). Social media outpaces print newspapers in the U.S. as news source | Pew Research Center. https://www.pewresearch.org/fact-tank/2018/12/10/social-media-outpaces-print-newspapers-in-the-u-s-as-a-news-source/

Sheridan, C. (2020). COVID-19 spurs wave of innovative diagnostics. Nature Biotechnology, 38(7), 769–772. https://doi.org/10.1038/s41587-020-0597-x

Sherman, E. (2020, February 21). Coronavirus impact: 94% of the Fortune 1000 are seeing supply chain disruptions | Fortune. Retrieved August 10, 2020, from https://fortune.com/2020/02/21/fortune-1000-coronavirus-china-supply-chain-impact/

Sherman, L. E., Michikyan, M., & Greenfield, P. M. (2013). The effects of text, audio, video, and in-person communication on bonding between friends. Cyberpsychology, 7(2). https://doi.org/10.5817/CP2013-2-3

Shih, S.-F., & Killeen, O. (2020). Increasing screen time during the coronavirus pandemic could be harmful to kids' eyesight. Retrieved July 31, 2020, from https://theconversation.com/increasing-screen-time-during-the-coronavirus-pandemic-could-be-harmful-to-kids-eyesight-138193

Shiri, R., & Falah-Hassani, K. (2015, February 15). Computer use and carpal tunnel syndrome: A meta-analysis. Journal of the Neurological Sciences, Vol. 349, pp. 15–19. https://doi.org/10.1016/j.jns.2014.12.037

Simon, C. (2020, April 16). Sleep problems becoming risk factor as pandemic continues – Harvard Gazette. Retrieved July 31, 2020, from https://news.harvard.edu/gazette/story/2020/04/sleep-problems-becoming-risk-factor-as-pandemic-continues/

Simonite, T. (2020). AI Uncovers a Potential Treatment for Covid-19 Patients | WIRED. Retrieved July 23, 2020, from https://www.wired.com/story/ai-uncovers-potential-treatment-covid-19-patients/

Singh, A. (2020). How AI is fighting COVID-19: the companies using intelligent tech to find new drugs -. Retrieved July 23, 2020, from https://pharmaphorum.com/views-analysis-digital/how-ai-is-fighting-covid-19-the-companies-using-intelligent-tech-to-find-new-drugs/

Smartbear. (n.d.). What is an API Endpoint? | SmartBear Software Resources. Retrieved August 7, 2020, from https://smartbear.com/learn/performance-monitoring/api-endpoints/

Smith, D. (2020, July 22). When will COVID end? The update on the race for a vaccine. Retrieved July 26, 2020, from https://www.cnet.com/how-to/when-will-covid-end-the-update-on-the-race-for-a-vaccine/

Smith, N. (2020, May 12). Video game industry giants have thrived in the covid-19 pandemic. Will the surge continue? - The Washington Post. Retrieved July 31, 2020, from https://www.washingtonpost.com/video-games/2020/05/12/video-game-industry-coronavirus/

Social Isolation Among Seniors: An Emerging Issue. (2004). British Columbia Ministry of
Health.
https://www.health.gov.bc.ca/library/publications/year/2004/Social_Isolation_A
mong_Seniors.pdf

Strasburger, V. C., Jordan, A. B., & Donnerstein, E. (2010). Health effects of media on children and adolescents. In Pediatrics (Vol. 125, Issue 4, pp.

756–767). American Academy of Pediatrics.
https://doi.org/10.1542/peds.2009-2563

Suciu, P. (2020, March 19). Fitness Goes To Social Media During COVID-19 Outbreak. Retrieved July 31, 2020, from https://www.forbes.com/sites/petersuciu/2020/03/19/fitness-goes-to-social-media-during-covid-19-outbreak/#545994cf38ea

Supply Chain Quarterly. (2020, June 3). Report: Product launches delayed due to Covid-19 | 2020-06-03 | CSCMP's Supply Chain Quarterly. Retrieved August 10, 2020, from https://www.supplychainquarterly.com/articles/3516-report-product-launches-delayed-due-to-covid-19

Survey: COVID-19 Affecting Patients' Access to Cancer Care American Cancer Society Cancer Action Network. (2020, April 15). American Cancer Society. https://www.fightcancer.org/releases/survey-covid-19-affecting-patients'-access-cancer-care

Swennen, G. R., Pottel, L., & Haers, P. E. (2020). Custom-made 3D-printed face masks in case of pandemic crisis situations with a lack of commercially available FFP2/3 masks. International Journal of Oral and Maxillofacial Surgery, 49(5), 673-677. doi:10.1016/j.ijom.2020.03.015

Tasker, J. P. (2020, April 28). House of Commons meeting virtually on a platform described as a "gold rush for cyber spies" | CBC News. Retrieved August 8, 2020, from https://www.cbc.ca/news/politics/zoom-house-of-commons-1.5546906

Tasnim, S., Hossain, M. M., & Mazumder, H. (2020). Impact of rumors or misinformation on coronavirus disease (COVID-19) in social media. Journal of Preventive Medicine & Public Health, 53(3), 171–174. https://doi.org/10.3961/jpmph.20.094

Technologies. (2020). Retrieved July 27, 2020, from https://www.medicago.com/en/technologies/

The Guardian. (2020a, April 4). NHL condemns "vile" racist outburst towards black player on Zoom chat | New York Rangers | The Guardian. Retrieved August 9, 2020, from https://www.theguardian.com/sport/2020/apr/04/k-andre-miller-racism-zoom-chat-hacker-nhl-new-york-rangers

The Guardian. (2020b, July 22). China hackers sought to steal coronavirus vaccine research, says US | China | The Guardian. Retrieved August 8, 2020, from https://www.theguardian.com/world/2020/jul/22/china-hackers-sought-to-steal-coronavirus-vaccine-research-says-us

131

The Health on the Net Foundation. (2019, November 5). HONcode.
https://www.hon.ch/HONcode/.

The Johns Hopkins University. (2020, June 30). Coronavirus Myths, Rumors
and Misinformation: Johns Hopkins Medicine. Coronavirus Myths,
Rumors and Misinformation.
https://www.hopkinsmedicine.org/coronavirus/articles/coronavirus-myths-
rumors-misinformation.html.

The Top 5 Practical Digital Health Technologies in the Fight Against COVID-
19: An Infographic - The Medical Futurist. (2020, May 7). The Medical
Futurist. https://medicalfuturist.com/the-top-5-practical-digital-health-
technologies-in-the-fight-against-covid-19-an-infographic/

Thompson, D. (2020). What Is "Contact Tracing" and How Does it Work?
Retrieved July 23, 2020, from
https://www.webmd.com/lung/news/20200504/what-is-contact-tracing-
and-how-does-it-work#1

Thompson, E. (2020, April 15). Federal government open to new law to fight
pandemic misinformation | CBC News. Retrieved August 8, 2020, from
https://www.cbc.ca/news/politics/covid-misinformation-disinformation-
law-1.5532325

Tkachenko, M. (2020). COVID-19 Accurately Diagnosed by AI Model.
Retrieved July 23, 2020, from https://www.genengnews.com/news/covid-
19-accurately-diagnosed-by-ai-model/

Tong, L. H., & Schwemle, B. L. (2002, April 3). Telework in the Federal
Government: Background, Policy, and Oversight. CRS Report for
Congress.
http://congressionalresearch.com/RL30863/document.php?study=Telework
+in+the+Federal+Government+Background+Policy+and+Oversight

Trivedi, T., Liu, J., Probst, J. C., Merchant, A., Jones, S., & Martin, A. B.
(2015). Obesity and obesity-related behaviors among rural and urban
adults in the USA. 15(4). https://doi.org/10.22605/RRH3267

Trussler, M., & Soroka, S. (2014). Consumer Demand for Cynical and Negative
News Frames. The International Journal of Press/Politics, 19(3), 360–379.
https://doi.org/10.1177/1940161214524832

Turbert, D. (2019, February 1). Nearsightedness: What Is Myopia? - American
Academy of Ophthalmology. Retrieved July 31, 2020, from
https://www.aao.org/eye-health/diseases/myopia-nearsightedness

U.S. Food and Drug Administration (FDA). (2020, July 06). N95 Respirators, Surgical Masks, and Face Masks. Retrieved July 14, 2020, from https://www.fda.gov/medical-devices/personal-protective-equipment-infection-control/n95-respirators-surgical-masks-and-face-masks

United Nations. (2020, April 14). COVID-19: Embracing digital government during the pandemic and beyond | Department of Economic and Social Affairs. Retrieved August 8, 2020, from https://www.un.org/development/desa/dpad/publication/un-desa-policy-brief-61-covid-19-embracing-digital-government-during-the-pandemic-and-beyond/

University of Waterloo. (2020). New AI technology will be used to improve contact tracing for COVID-19 | Waterloo Stories | University of Waterloo. Retrieved July 23, 2020, from https://uwaterloo.ca/stories/news/new-ai-technology-will-be-used-improve-contact-tracing-covid

Ura, T., Okuda, K., & Shimada, M. (2014). Developments in Viral Vector-Based Vaccines. Vaccines, 2(3), 624-641. doi:10.3390/vaccines2030624

Vad, V. (2020, January 9). Is Poor Posture Causing Your Back Pain? Retrieved July 31, 2020, from https://www.spine-health.com/blog/poor-posture-causing-your-back-pain

Van der Sande, M., Teunis, P., & Sabel, R. (2008). Professional and Home-Made Face Masks Reduce Exposure to Respiratory Infections among the General Population. PLoS ONE, 3(7). doi:10.1371/journal.pone.0002618

Ventura Orthopedics Staff. (2018, July 3). How To Prevent Smartphone Hand Pain - Ventura Orthopedics. Retrieved July 31, 2020, from https://venturaortho.com/how-to-prevent-smartphone-hand-pain/

Vilhelmson, B., & Thulin, E. (2016). Who and where are the flexible workers? Exploring the current diffusion of telework in Sweden. New Technology, Work and Employment, 31(1), 77–96. https://doi.org/10.1111/ntwe.12060

Vogels, E. A. (2020, April 30). What Americans are doing online during COVID-19 | Pew Research Center. Retrieved July 31, 2020, from https://www.pewresearch.org/fact-tank/2020/04/30/from-virtual-parties-to-ordering-food-how-americans-are-using-the-internet-during-covid-19/

Wakabayashi, D., Nicas, J., Lohr, S., & Isaac, M. (2020, March 23). Big Tech Could Emerge From Coronavirus Crisis Stronger Than Ever - The New York Times. Retrieved August 10, 2020, from https://www.nytimes.com/2020/03/23/technology/coronavirus-facebook-amazon-youtube.html

Walsh, L. (2020, April 9). Next Problem: Delayed Product Launches –
Channelnomics. Retrieved August 10, 2020, from
https://channelnomics.com/2020/04/09/next-problem-delayed-product-
launches/

Waltz, E. (2020). How Computer Scientists Are Trying to Predict the
Coronavirus's Next Moves - IEEE Spectrum. Retrieved July 23, 2020,
from https://spectrum.ieee.org/the-human-
os/biomedical/devices/predicting-the-coronavirus-next-moves

Warren, M. (2020, April 10). Health-care workers at two Toronto hospitals told
they'll soon be reusing decontaminated N95 masks. Retrieved August 7,
2020, from https://www.thestar.com/news/gta/2020/04/10/health-care-
workers-at-two-toronto-hospitals-told-theyll-soon-be-reusing-
decontaminated-n95-masks.html

Watson, A. (2020, May 5). Fake News - Statistics & Facts | Statista. Retrieved
August 8, 2020, from https://www.statista.com/topics/3251/fake-
news/#dossierSummary

Weikle, B. (2020, April 2). Long waits for COVID-19 test results reveal
Canada's shortage of lab workers. CBC News.
https://www.cbc.ca/news/health/medical-laboratory-worker-shortage-
1.5517908

Wilcox, C., Mallos, N. J., Leonard, G. H., Rodriguez, A., & Hardesty, B. D.
(2016). Using expert elicitation to estimate the impacts of plastic pollution
on marine wildlife. Marine Policy, 65, 107-114.
doi:10.1016/j.marpol.2015.10.014

Wittbold, K. A., Carroll, C., Iansiti, M., Zhang, H. P., & Landman, A. B.
(2020). How Hospitals Are Using AI to Battle Covid-19. Retrieved July
23, 2020, from https://hbr.org/2020/04/how-hospitals-are-using-ai-to-
battle-covid-19

Wong, H., Moore, K., Angstman, K. B., & Garrison, G. M. (2019). Impact of
rural address and distance from clinic on depression outcomes within a
primary care medical home practice. BMC Family Practice, 20(1), 123.
https://doi.org/10.1186/s12875-019-1015-7

Wood, L. (2020, May 7). COVID-19 Outbreak: Video Conferencing Demand
Rises due to Social-Distancing - ResearchAndMarkets.com | Business
Wire. Retrieved August 10, 2020, from
https://www.businesswire.com/news/home/20200507005631/en/COVID-
19-Outbreak-Video-Conferencing-Demand-Rises-due

World Health Organization (WHO). (2011, October 12). DNA vaccines. Retrieved July 26, 2020, from https://www.who.int/biologicals/areas/vaccines/dna/en/

World Health Organization (WHO). (2020a, July 30). Coronavirus Disease (COVID-19) - events as they happen. Retrieved August 10, 2020, from https://www.who.int/emergencies/diseases/novel-coronavirus-2019/events-as-they-happen

World Health Organization (WHO). (2020b, March 19). Infection prevention and control during health care when novel coronavirus (nCoV) infection is suspected. Retrieved July 16, 2020, from https://www.who.int/publications/i/item/infection-prevention-and-control-during-health-care-when-novel-coronavirus-(ncov)-infection-is-suspected-20200125

World Health Organization (WHO). (2020c, March 3). Shortage of personal protective equipment endangering health workers worldwide. Retrieved July 13, 2020, from https://www.who.int/news-room/detail/03-03-2020-shortage-of-personal-protective-equipment-endangering-health-workers-worldwide

World Health Organization (WHO). (2020d, March 8). Media Statement: Knowing the risks for COVID-19. World Health Organization. Retrieved August 10, 2020 from https://www.who.int/indonesia/news/detail/08-03-2020-knowing-the-risk-for-covid-19

World Health Organization (WHO). (2020e, April 23). WHO reports fivefold increase in cyber attacks, urges vigilance. Retrieved August 8, 2020, from https://www.who.int/news-room/detail/23-04-2020-who-reports-fivefold-increase-in-cyber-attacks-urges-vigilance

Worldometer. (2020). Coronavirus Cases. Worldometer, 1–22. https://doi.org/10.1101/2020.01.23.20018549V2

Wu, D. (2020, March 27). IPhone Makers Look Beyond China in Supply-Chain Rethink - Bloomberg. Retrieved August 10, 2020, from https://www.bloomberg.com/news/articles/2020-03-27/iphone-makers-look-beyond-china-in-supply-chain-rethink

Wu, Jianguo, et al. "Detection and Analysis of Nucleic Acid in Various Biological Samples of COVID-19 Patients." Travel Medicine and Infectious Disease, 2020, p. 101673., doi:10.1016/j.tmaid.2020.101673.

Yasinski, Emma. "Scientists Scrutinize New Coronavirus Genome for Answers." The Scientist Magazine®, 2020, www.the-scientist.com/news-opinion/scientist-scrutinize-new-coronavirus-genome-for-answers-67006.

Zanni, T. (2020, April 21). Tech COVID-19 supply chain disruption - KPMG Global. Retrieved August 10, 2020, from https://home.kpmg/xx/en/blogs/home/posts/2020/04/technology-supply-chain-disruption.html

www.ingramcontent.com/pod-product-compliance
Lightning Source LLC
Chambersburg PA
CBHW031811190326
41518CB00006B/294